ASSESSING BACTERIAL GROWTH PO[TENTIAL IN SEAWATER] REVERSE OSMOSIS PRET[REATMENT:] METHOD DEVELOPMENT AN[D APPLICATION]

ALMOTASEMBELLAH M. J. ABUSHABAN

ASSESSING BACTERIAL GROWTH POTENTIAL IN SEAWATER REVERSE OSMOSIS PRETREATMENT: METHOD DEVELOPMENT AND APPLICATIONS

DISSERTATION

Submitted in fulfilment of the requirements of
the Board for Doctorates of Delft University of Technology
and
of the Academic Board of the IHE Delft
Institute for Water Education
for
the Degree of DOCTOR
to be defended in public on
3rd, December 2019, at 15:00 hours
in Delft, the Netherlands

by

Almotasembellah Mustafa Jawdat ABUSHABAN
Master of Science in Municipal Water and Infrastructure
UNESCO-IHE, Institute for Water Education, the Netherlands
born in Kuwait, Kuwait

This dissertation has been approved by the
promotor: Prof. dr. ir. M.D. Kennedy and
copromotor: Dr. ir. S.G. Salinas Rodriguez

Composition of the doctoral committee:
Rector Magnificus TU Delft   Chairman
Rector IHE Delft   Vice-Chairman
Prof. dr. ir. M. D. Kennedy   IHE Delft/ TU Delft, promotor
Dr. ir. S.G. Salinas Rodriguez   IHE Delft, copromotor

Independent members:
Prof. dr. P.A. Davies   University of Birmingham, UK
Prof. dr. ir. L.C. Rietveld   Delft University of Technology
Prof. dr. ir. W.G.J. van der Meer   University of Twente
Dr. E. Prest   PWNT, Netherlands
Prof. dr. M.E. McClain   TU Delft/IHE Delft, reserve member

This research was conducted under the auspices of the Graduate School for Socio-
Economic and Natural Sciences of the Environment (SENSE)

CRC Press/Balkema is an imprint of the Taylor & Francis Group, an informa business
© 2019, Almotasembellah M. J. Abushaban

*Although all care is taken to ensure integrity and the quality of this publication and the
information herein, no responsibility is assumed by the publishers, the author nor IHE Delft for
any damage to the property or persons as a result of operation or use of this publication and/or
the information contained herein.*

*A pdf version of this work will be made available as Open Access via
http://repository.tudelft.nl/ihe This version is licensed under the Creative Commons Attribution-
Non Commercial 4.0 International License, http://creativecommons.org/licenses/by-nc/4.0/*

Published by:
CRC Press/Balkema
Schipholweg 107C, 2316 XC, Leiden, the Netherlands
Pub.NL@taylorandfrancis.com
www.crcpress.com – www.taylorandfrancis.com
ISBN 978-0-367-89906-6

To my family all over the world …

ACKNOWLEDGMENTS

I would like to express my gratitude and appreciation to all individuals who contributed to the success of this thesis. All things are possible through collaboration.

Foremost, I would like to express my sincere gratitude to my promoter, Prof. Maria D. Kennedy for giving me the opportunity to carry on my academic career and for her strong support, wise guidance, encouragement and brilliant advice throughout the research study. I am also thankful to my supervisors, Sergio Salinas- Rodriguez for his support, guidance, assistance, availability, time, and patience and to Prof. Jan C. Schippers for teaching me view things in a broad way and his advice and critical discussions and suggestions.

Thanks are due to Promega for co-funding my research project at IHE-Delft, the Netherlands and for funding my field research. Thanks go to Zhong Yu, Duddy Oyib, Brigitta Saul, Said Goueli, Subhanjan Mondal, David Grasso, Tim Deschaines, Patrick van der Velde and Hemanth Shenoi. I would also like to thank Grundfos for co-funding my PhD research and for the knowledge I have gained during working on scaling of reverse osmosis and designing a scale guard unit. . In particular, I thank Victor Augusto Yangali Quintanilla, Wilbert van de Ven and Vincent Groenendijk.

This work involved pilot and full-scale plants testing in different countries including Australia, Oman, United Arab Emirates and the Netherlands. This work would not have been possible without the cooperation of SUEZ (Delia Pastorelli, Sophie Bertrand, Emmanuelle Filloux, Remy Caball, Elias Felix and Daniel Wrabel), Veolia (David Cohen, Scott Murphy and Diannne Turner), ENGie (Ayoub Al Azri, Younis Albalushi), Zeeschelp (Marco Dubbeldam, Bernd van Broekhoven and Hanno), Applitek (Hannes Hoste) and IHE Delft lab staff (Fred Kuis, Peter Heerings, Frank Wiegman, Berend Lolkema, Ferdi Battes and Lyzette Robbemont).

I was happy to supervise six master students and one bachelor student during my research project. I would like to express my deepest appreciation to all of them for their contribution to this research work (Almohanad Abusultan, Anjar Prabowo, Leon Ramattan, Muhammad Nasir Mangal, Chidiebere Samuel Ebere, Moses Kapala and Maud Salvaresi).

I would like to thank Prof. Hans Vrouwenvelder, Bastiaan Blankert, Loreen Villacorte and Nirajan Dhakal for their critical comments and insightful discussions, which also contributed to the success of the research.

My thanks to all my friends and colleagues for making my academic life in Delft a social experience. My gratitude goes to (in alphabetic order); Ahmad Mahmoud, Ahmed Elghandour, Alida Alves Beloqui, Angelica Rada, Bianca Wassenaar, Chris Metzker, Conny Hoek, Ellen de Kok, Fiona Zakaria, Gabriela Cuadrado Quesada, Hesham Elmilady, Iosif Skolluos, Jolanda Boots, Joyabrata Mal, Lea Tan, Mamoun Althuluth, Maria Salingay, Maria Reyes, Marmar Ahmed, Mohan Radakrishnan, Mohaned Sousi, Mohanad Abunada, Muhammed Dikman Maheng, Musad Aklan, Nasir Mangal, Pia Legerstee-de Vries, Shrutika Wadgaonkar, Rohan Jain, Ronald Mollinger, Ruknul Ferdous, Selvi Pransiska, Shahnoor Hasan, Shakeel Hayat, Suzanne Lander, Taha Al Washali, Teju Madapura Eregowda, Yuli Ekowati, Yousef Albuhaisi.

I am extremely grateful to my parents for their endless support, love, prayers and care that kept me going although I was miles away from them, and for their sacrifices during my stay abroad. Special thanks goes to my wife and daughter for their love, understanding, motivation, prayers and continuing support during my PhD research project. I am also very thankful to my brothers, sisters, brothers-in-law and sisters-in-law for believing in me and for their support, prayers and wishes. Thanks to all my friends all over the world for their wishes.

Motasem Abushaban

December 2019
Delft, Netherlands

SUMMARY

Seawater desalination is expanding rapidly and the current global installed capacity has reached over 95 million cubic meters per day. Since 2000, reverse osmosis (RO) has been the most dominant technology used in seawater desalination because of the lower energy consumption, lower cost and smaller footprint compared to thermal systems. Currently, approximately, two-thirds of desalinated seawater is produced by seawater reverse osmosis (SWRO) membranes in more than 4,300 SWRO desalination plants globally.

Membrane fouling is the main challenge that SWRO systems face during operation. Pre-treatment is commonly applied to improve water quality prior to reverse osmosis (RO), and minimize/mitigate fouling issues in SWRO systems. Fouling caused by colloidal particles is generally well controlled through SWRO pre-treatment, with SWRO feed water achieving the target Silt Density Index (SDI) < 5 %/min in most cases. However, many SWRO desalination plants still struggle to control biological and organic fouling as there are no standard methods to monitor these types of fouling in desalination plants. Biological fouling results from microbial growth in membrane systems, which may lead to operational problems such as increased head loss across feed spacers in spiral wound elements and decreased permeability of RO membranes. Biofilm formation in SWRO is inevitable if the feed water supports significant bacterial growth due to the presence of easily biodegradable (dissolved) nutrients.

The use of Assimilable Organic Carbon (AOC) and Bacterial Growth Potential (BGP) methods to monitor biofouling potential in SWRO feed water, has gained interest as high levels of AOC/BGP are directly linked to biofilm formation and thus loss of performance in membrane processes. However, the relationship between these methods and biofouling development in full-scale plants has not yet been demonstrated for SWRO plants. A preliminary guideline value (<1 µg-C/L as acetate) was proposed for freshwater RO plants, to avoid biofouling.

Existing AOC methods for seawater make use of a single pure strain of marine bacteria (e.g. *Vibrio fischeri* and *Vibrio harveyi*) as inoculum, which may not reflect the real carbon utilization of indigenous microorganisms in seawater and thus may underestimate the nutrient concentration of seawater. To overcome this problem, bacterial growth potential (BGP) in seawater is proposed using an indigenous microbial consortium to provide more predictive information regarding the nutrient concentration, compared with the use of a pure bacterial strain (as inoculum).

To monitor the growth of an indigenous microbial consortium, fast and accurate bacterial enumeration methods are required. Adenosine Triphosphate (ATP) has gained interest because it is accurate, rapid, can detect both cultivable and uncultivable microorganisms, is easy to perform and most importantly, it is based on the activity of microorganisms. However, with existing ATP methods, ATP cannot be measured accurately at low levels in high saline environments like seawater, due to the interference of salt with the luciferin/luciferase reaction, which until now has inhibited light production and ATP measurement in seawater.

The main goal of this research was to develop, test and validate a new method to measure bacterial growth potential (BGP) in seawater, based on microbial ATP. For this purpose, a method was developed to measure microbial ATP in seawater without interference from the seawater matrix. Subsequently, the new microbial ATP method was used to monitor BGP in seawater making use of an indigenous microbial consortium. Thereafter, the BGP and ATP methods as well as other fouling indices were used to assess the removal of biological/organic fouling potential, through the pre-treatment processes of several pilot and full-scale SWRO desalination plants. Finally, the correlation between BGP in SWRO feed water and real time fouling development in several SWRO plants was investigated.

Firstly, a direct method to measure microbial ATP in seawater was developed, which involved extraction of ATP directly from biomass in seawater, followed by ATP detection using new reagents especially developed for seawater. The ATP-direct method is fast (< 5 min) and sensitive (LOD of 0.3 ng-ATP/L). However, the pH and iron concentration (added as coagulant) in seawater negatively affect the luminescence signal. To overcome this effect, a calibration line (with the same seawater matrix as the actual seawater sample) is required as seawater characteristics (such as; salinity, pH and iron concentration) may change along pre-treatment processes. To eliminate the interference of the seawater matrix during ATP measurement, a filtration-based ATP method was also developed, whereby microorganisms are captured on a membrane surface. The filtration step eliminates all matrix effects and increases the sensitivity of the method (LOD ≤ 0.06 ng-ATP/L, depending on the filtered sample volume). Several variables that might affect the performance of the ATP-filtration method were tested and optimized, including the filter pore size, rinsing of free ATP from the filter holder, and the effect of the seawater sample volume. Microbial ATP concentrations measured using the ATP-filtration method were comparable to concentrations measured using the ATP-direct method in seawater. Moreover, microbial ATP concentrations in seawater measured with the ATP-filtration method correlated ($R^2 = 0.72$, n = 100) with the intact cell concentration measured by flow cytometry.

The new ATP-based methods was used to measure BGP using an indigenous microbial consortium. Each step in the BGP protocol was studied and optimized, including bacterial inactivation, bacterial inoculation, incubation and bacterial growth monitoring. The limit of detection of the BGP method is 13 µg C-glucose/L ($\sim$ 10 µg C-acetate/L) measured in artificial seawater (blank). The bacterial yield in seawater ranged between 1 and 1.5 ng-ATP/µg C-glucose, and was tested at five different locations in Australia and the Middle East. BGP and ATP concentrations in the North Sea were monitored for 12 months, with seasonal variations ranging from 45 µg glucose-C/L in the winter to 385 µg glucose-C/L in the spring and from 25 ng-ATP/L in the winter to 1,035 ng-ATP/L in the spring, respectively.

Subsequently, the ATP and BGP methods were applied to monitor the pre-treatment processes of several full-scale SWRO desalination plants located in Australia and the Middle East. The highest removal of microbial ATP (50 – 95 %) and BGP (55 – 70 %) was achieved with dual media filtration (DMF) with inline coagulation (0.8 – 3.6 mg Fe^{3+}/L). The removal of BGP in DMF combined with inline coagulation (0.8 mg Fe^{3+}/L) was similar to that observed removal in dissolved air floatation (DAF) combined with ultrafiltration (UF).

The microbial ATP and BGP methods were also applied, together with other fouling indicators, to compare the reduction in fouling potential achieved at two full-scale SWRO desalination plants with different seawater intake sources and pre-treatment systems. The pre-treatment in the first plant included inline coagulation ($0.7 - 1.7$ mg-Fe^{3+}/L) and two stages of DMF, while the pre-treatment of the second plant included DAF, inline coagulation ($0.3 - 1.5$ mg-Fe^{3+}/L) and two stages of DMF. The indicators and parameters monitored in the plants included particulate fouling indices (SDI and modified fouling index (MFI)), biological/organic indices (BGP, orthophosphate, organic fractions measured by liquid chromatography organic carbon detection (LC-OCD) and total organic carbon (TOC)), microbial ATP, and turbidity.

In both SWRO plants, DMF combined with inline coagulation showed more than 75 % of particulate fouling and microbial ATP. However, the removal of organic/biological fouling potential observed in DMF combined with inline coagulation was much lower (< 40 %). The poor removal in these plants was attributed to the low biological activity in the DMF because chlorine (applied at the intake) was not neutralized prior to DMF. In addition, SWRO brine was used (instead of filtered seawater) to backwash the DMF. Both the presence of chlorine in DMF feed water as well as brine backwashing may stress/damage the active biofilm layer on the filter media, and this was supported by ATP and BGP measurements during DMF operation and backwashing. Neutralizing chlorine before media filtration will support the development of an active biofilm layer on the media, capable of degrading/removing easily biodegradable organic matter in the feed water. Additionally, backwashing the media filters with filtered seawater may also significantly improve the biological activity in the media filters as well as biodegradation of organic matter.

The removal of particulate/biological/organic fouling potential in the second stage DMF was always lower than in the first stage which could be attributed to the absent of coagulant prior to the second stage of DMF and/or shorter contact time compared with the first stage of DMF. In both SWRO plants, the BGP increased prior to the SWRO membranes due to antiscalant addition. Comparing the overall removal within the pre-treatment of the two SWRO desalination plants, a positive impact of DAF on the removal of biological/organic fouling potential was observed. The additional removal of BGP, orthophosphate, CDOC and biopolymers in the DAF system ($1 - 5$ mg-Fe^{3+}/L) was 52 %, 68 %, 8 % and 25 %, respectively.

Finally, an attempt was made to investigate if any correlation exists between the BGP of SWRO feed water and the chemical cleaning frequency (as a surrogate parameter for biofouling) in SWRO plants. Investigating the existence of such a correlation is complicated by several factors. Firstly, several types of fouling (scaling, particulate and organic/biofouling) may occur simultaneously. Secondly, the widespread intermittent use of non-oxidizing biocides to combat biofouling in full-scale SWRO facilities makes establishing any real correlation between the BGP of SWRO feed water and membrane performance very difficult. Thirdly, in order to establish a robust correlation, a large number of SWRO desalination plants in different parts of the world need to be monitored for longer periods of time. Despite these limitations, the correlation between the BGP of SWRO feed water and the chemical cleaning frequency was investigated in four full-scale SWRO desalination plants and the results showed that a higher BGP in SWRO feed water corresponded to a higher chemical cleaning frequency in the SWRO system.

Thereafter, the correlation between the BGP of SWRO feed water and the normalized pressure drop was investigated in a full-scale SWRO desalination plant monitored for a period of 5 months. A higher normalized pressure drop was observed in the SWRO membrane system with higher levels of BGP in the SWRO feed water (from 100 to– 950 µg-C/L), indicating that the BGP method may be a useful indicator of biological fouling in SWRO systems. Based on BGP measurements performed in five full-scale SWRO plants, a tentative threshold concentration of BGP (< 70 µg/L) is proposed for SWRO feed water in order to ensure a chemical cleaning frequency of once/year or lower. However, to verify the level recommended above, more data needs to be collected and many more SWRO plants need to be monitored for longer periods of time under different operating conditions.

Overall, a new ATP-based BGP method was developed, using an indigenous microbial consortium, to monitor and assess the removal of biological/organic fouling potential through SWRO pre-treatment systems such as DAF, DMF and UF. The methods can be used to optimize the operation of media filtration (i.e. contact time, filter backwashing, chlorination/dechlorination, coagulation) as wells as DAF and UF systems to achieve a low BGP and organic fouling potential in SWRO feed water.

Finally, further developments of the new ATP-based BGP method are still needed. Lowering the LOD of the method even further is recommended to broaden it applicability to brackish water and seawater with low biofouling potential (such as SWRO plants fed with seawater from beach wells). In addition, a threshold value of BGP in SWRO feed water is required as a guideline for full-scale plants to assess the efficiency of their pre-treatment systems. In addition, the BGP protocol is rather complicated, and developing both online ATP and BGP analysers for seawater would be beneficial for SWRO plants as well as applications in ballast and swimming pool water.

SAMENVATTING

Zeewaterontzilting neemt in steeds drastischere mate toe, met een huidige geïnstalleerde capaciteit van ongeveer 95 miljoen kubieke meter per dag. Sinds 2,000 is omgekeerde osmose (RO) de meest dominante technologie voor ontzilting. Momenteel wordt ongeveer twee derde van het ontzilte zeewater geproduceerd door middel van zeewater omgekeerde osmose (SWRO) membranen in meer dan 4,300 SWRO ontziltingsinstallaties wereldwijd.

Membraanvervuiling wordt gezien als de grootste uitdaging waarmee de werking van SWRO systemen mee geconfronteerd worden. Nog voor de omgekeerde osmose (RO), wordt vaak een voorzuivering toegepast om de waterkwaliteit te verbeteren, en om zo het vervuilingsprobleem in SWRO systemen tegen te gaan/te minimaliseren. Vervuiling die veroorzaakt wordt door colloïdale deeltjes, kan goed gereguleerd worden door de SWRO voorzuivering door het produceren van acceptabele SWRO voedingswaterkwaliteit (met zilt dichtheid index, SDI < 5 %/min). Veel SWRO ontziltingsinstallaties kampen echter met biologische- en organische vervuilingsproblemen, en hier zijn geen standaard methoden voor om deze vervuilingstypen te observeren en te controleren. Biologische vervuiling komt voort uit microbiële groei in membraansystemen wat kan leiden tot operationele problemen, zoals een toename van drukverlies in de filter voorruimte van spiraal-gewikkelde elementen, en een afname van de doorlatendheid van RO membranen. De vorming van biofilm in SWRO is onvermijdelijk wanneer het voedingswater een aanzienlijke bacteriële groei ondersteunt door de aanwezigheid van gemakkelijk afbreekbare opgeloste voedingsstoffen. Daarom heeft het gebruik van methoden voor groeipotentieel, zoals assimilable organic carbon (AOC) veel interesse gewonnen, aangezien ze direct gerelateerd kunnen zijn aan de vorming van biofilm op RO membraan. De verhouding tussen deze methoden en de ontwikkeling van vervuiling (biofouling) in uitgebreide installaties is nog niet vastgesteld. De bestaande AOC methoden in zeewater gebruiken een pure soort zoals inoculum. Het kan zijn dat deze pure soort het gebruik van koolstof van inheemse micro-organismen in zeewater niet goed reflecteert door het gebrek aan interacties tussen verschillende bacteriën, waardoor de nutriëntenconcentratie in zeewater onderschat kan worden. Daarom kan het meten van bacteriële groeipotentie

(BGP) in zeewater door middel van een inheems microbieel consortium meer voorspellende informatie verstrekken dan het gebruik van een pure bacteriële soort zoals inoculatie.

Om de groei van een inheems microbieel consortium te observeren zijn ook snelle en accurate bacteriële opsommingsmethoden nodig. Adenosine Triphosphate (ATP) heeft een toegenomen interesse gewonnen omdat het nauwkeurig en snel is, het zowel cultiveerbare – als niet-cultiveerbare micro-organismen kan detecteren, het eenvoudig uit te voeren is en het, bovenal, gebaseerd is op de activiteitsgraad van de micro-organismen. Met de huidige ATP methoden kan ATP echter niet nauwkeurig gemeten worden bij lage niveaus in omgevingen met een hoog zoutgehalte, zoals zeewater, door de verstoring van zout met de luciferine/luciferase reactie, welke tot op heden lichtproductie en ATP meting heeft belemmerd.

Het hoofddoel van dit onderzoek was het ontwikkelen, testen en valideren van een nieuwe methode waarmee bacteriële groeipotentie (BGP) in zeewater, op basis van microbiële ATP, mee gemeten kan worden. De methode is bedoeld om de verwijdering van biologische/organische vervuilingspotentie door de voorzuivering van SWRO systemen te observeren en te beoordelen, met als doel de biologische/organische vervuiling in SWRO membranen te controleren. Te dien einde was een methode ontwikkeld om microbiële ATP in zeewater, zonder enige verstoring van de zeewatermatrix, te meten. Vervolgens werd de nieuw-ontwikkelde microbiële ATP methode gebruikt om BGP in zeewater te observeren door middel van een inheems microbieel consortium. Daarna werd het verband tussen BGP in RO voedingswater en de werkelijke vervuilingsontwikkeling onderzocht. Tenslotte werd de verwijdering van biologische/organische vervuilingspotentie beoordeeld door verschillende SWRO voorzuiveringsprocessen bij verscheidene pilot- en uitgebreide ontziltingsinstallaties.

Allereerst werd een directe methode, waarmee microbiële ATP in zeewater gemeten kan worden, ontwikkeld. Deze methode omvat de onttrekking van ATP direct uit biomassa in zeewater, gevolgd door de vaststelling hiervan, door middel van het gebruik van nieuwe reagentia die speciaal voor zeewater ontwikkeld zijn. De directe ATP-methode is direct, snel (< 5 min) en sensitief (LOD of 0.3 ng-ATP/L). Toch toonde pH van het water en ijzerconcentratie (welke toegevoegd kan worden als coagulant), die zich in het zeewater bevinden, een negatief effect op het geproduceerde luminescentiesignaal. Om dit effect te overstijgen is daarom een merkstreep met een vergelijkbare zeewatermatrix tot het werkelijke zeewatermonster nodig, aangezien zeewaterkenmerken (zoals saliniteit, pH en ijzerconcentratie) gedurende de voorzuiveringsprocessen kunnen veranderen. Om de verstoring van de zeewatermatrix met de ATP meting te elimineren, werd een filtratie-gebaseerde methode geïntroduceerd om de micro-organismen op een membraanoppervlak te vangen. Deze filtratiestap verhoogde ook de sensitiviteit van de methode (LOD ≤ 0.06 ng-ATP/L, afhankelijk van het volume van het gefilterde monster). De microbiële ATP-concentratie, die gemeten werd met de ATP-filtratiemethode, was vergelijkbaar met de concentratie die gemeten werd met behulp van de ATP-direct methode in zeewater. Bovendien kwam de microbiële ATP-concentratie, gemeten met de ATP-filtratiemethode ($R^2 = 0.72$, $n = 100$), overeen met de intact gebleven cel-concentratie die gemeten werd door flow-cytometrie. De methoden werden toegepast om de microbiële ATP-concentratie te observeren gedurende de voorzuiveringsprocessen van drie uitgebreide SWRO ontziltingsinstallaties in Australië en in het Midden-Oosten, waar de verwijdering van microbiële ATP met het hoogste percentage (50 – 95 %) werd geobserveerd in filtratiemedia.

Ten tweede werd een nieuw, op ATP-gebaseerde methode ontwikkeld om de BGP te meten door middel van een inheems microbieel consortium. De detectielimiet van de BGP is 13 μg C-glucose/L (~ 10 μg C-acetaat/L). Een verlaging van de LOD zou ideaal zijn voor het meten van lage BGP in de SWRO-voeding, voornamelijk gedurende de winter. Bio-vervuiling wordt echter niet verondersteld bij lage watertemperaturen met een lage BGP. De bacteriële opbrengst varieerde tussen de 1 en 1.5 ng-ATP/μg C-glucose, getest op vijf verschillende locaties wereldwijd. De methode werd toegepast om BGP in de Noordzee te observeren in een periode van 12 maanden. Hierbij werden seizoensgebonden variaties geobserveerd, variërend van 45 μg glucose-C/L gedurende de winter en 385 μg glucose-C/L gedurende de lente. Bovendien werd BGP waargenomen tijdens de voorzuiveringsseries van verscheidene complete SWRO ontziltingsinstallaties met verschillende voorzuiveringsprocessen. Duale mediafiltratie (DMF), gecombineerd met inline coagulatie (0.8 – 3.6 mg Fe^{3+}/L), toonde het hoogste percentage BGP verwijdering (55 – 70 %) in twee SWRO installaties in het Midden-Oosten en Australië.

De verwijdering van BGP en hydrofiele opgeloste organische koolstof (CDOC) in opgeloste lucht flotatie (DAF), gecombineerd met ultrafiltratie (UF), was vergelijkbaar met de verwijdering in DMF (contacttijd = 5 min.) in combinatie met inline coagulatie $(0.8 \text{ mg Fe}^{3+}/L)$.

Ten derde werden de ontwikkelde microbiële ATP- en BGP-methoden toegepast, samen met andere vervuilingsindicatoren, om de voorzuivering van twee complete SWRO ontziltingsinstallaties met verschillende bronnen van zeewaterinname (Golf van Oman en de Perzische Golf) te beoordelen. De voorzuivering van de eerste installatie omvat inline coagulatie $(1.3 \text{ mg-Fe}^{3+}/L)$ en twee fasen van DMF, terwijl de voorzuivering van de tweede installatie DAF, inline coagulatie $(0.35 \text{ mg-Fe}^{3+}/L)$ en twee fasen van DMF omvat. De waargenomen indicatoren en parameters bevatten deeltjesvervuiling-indexen (SDI en gemodificeerde vervuilingsindex (MFI)), biologische/organische indexen (BGP, orthofosfaat concentratie liquide chromatografie organische detectie (LC-OCD) en totale organische koolstof (TOC), microbiële ATP, vertroebeling, en totale ijzerconcentratie.

In beide SWRO installaties toonde DMF, gecombineerd met inline coagulatie, een hoog percentage verwijdering (> 75 %) van deeltjesvervuilingspotentie en microbiële ATP aan. Toch werd een lager percentage verwijdering van biologisch organische vervuilingspotentie (< 40 %) waargenomen in DMF, gecombineerd met inline coagulatie. Het lage percentage verwijdering van biologisch/organische vervuilingspotentie kon worden toegeschreven aan de lage biologische activiteit in DMF, door de frequente chlorering (wekelijks) en het gebruik van zeewater met een hoog zoutgehalte (SWRO pekel) voor filter terugspoeling, welke hoogstwaarschijnlijk een osmotische drukgolf op de biofilmlaag van de media filter veroorzaakte. De verwijdering van deeltjes/biologische/organische vervuilingspotentie in de tweede fase DMF was altijd lager dan in de eerste fase DMF, wat toegeschreven kon worden aan de afwezigheid van de dosis coagulant voorafgaand aan de tweede fase van DMF en/of een kortere contacttijd (3 min.) vergeleken met DMF1 (5 min.). In beide SWRO installaties werd een aanzienlijke potentie van organische/biologische vervuiling waargenomen in het SWRO voedingswater door de toevoeging van antiscalant.

Tijdens het vergelijken van de globale verwijdering in de voorzuivering van de twee SWRO ontziltingsinstallaties, werd het duidelijk dat een positieve impact van DAF aangetoond kon worden in het verbeteren van de verwijdering van biologische/organische vervuilingspotentie. De aanvullende verwijdering van BGP, orthofosfaat, CDOC en biopolymeren in DAF systemen was respectievelijk 52%, 68%, 8% en 25%. De hoge verwijdering van BGP en biopolymeren in DAF systemen kan toegeschreven worden aan de toevoeging van de dosis coagulant (1.5 mg- Fe^{3+}/L) in DAF.

Tenslotte werd geprobeerd om de correlatie tussen BGP van SWRO voedingswater en bio-vervuiling in SWRO te onderzoeken. Het onderzoeken van zo'n correlatie wordt bemoeilijkt door verscheidene factoren. Ten eerste kunnen verschillende typen vervuiling (schalings-, deeltjes- en organische/bio-vervuiling) tegelijkertijd voorkomen. Ten tweede bemoeilijkt het grootschalige-, intermitterende gebruik van niet-oxiderende biociden, die gebruikt worden bij het bestrijden van bio-vervuiling in complete SWRO faciliteiten, het tot stand brengen van enige echte correlatie tussen de BGP van SWRO voedingswater en membraan-prestatie. Ten derde moet een groot aantal SWRO installaties in verschillende delen van de wereld voor langere tijd gecontroleerd worden onder verschillende operationele omstandigheden.

Daarom werd aanvankelijk de relatie tussen BGP van SWRO voedingswater en de chemische reinigingsfrequentie geverifieerd in drie grote SWRO ontziltingsinstallaties, waarbij duidelijk werd dat een hogere BGP in SWRO voedingswater correspondeerde met een hogere chemische reinigingsfrequentie. Vervolgens werd de correlatie tussen BGP in SWRO voedingswater en de genominaliseerde drukdaling onderzocht in een complete SWRO ontziltingsinstallatie, en daarnaast ook voor vijf maanden geobserveerd. Een hogere genominaliseerde drukdaling in het SWRO membraansysteem werd waargenomen bij een hogere BGP in het SWRO voedingswater, wat de toepasbaarheid van BGP als biologische vervuilingsindicator in het SWRO systeem aantoonde. Echter moet meer data verzameld worden en meer SWRO installaties moeten voor langere periodes gecontroleerd worden onder verschillende operationele omstandigheden.

Globaal gezien werd een nieuwe ATP-gebaseerde BGP methode ontwikkeld door middel van een inheems microbieel consortium dat gebruikt kan worden bij het observeren en beoordelen van biologische/organische vervuilingspotentie gedurende de SWRO voorzuivering. De ontwikkeling van deze methode richt zich op het controleren van biologische/organische vervuiling van het SWRO systeem door correctieve maatregelen uit te voeren in een vroeg stadium van de voorzuivering. Deze studie benadrukt de verwijderingsefficiëntie van verschillende voorzuiveringsprocessen, waaronder DAF, DMF en UF, en toonde aan dat de nieuw-ontwikkelde methoden gebruikt kunnen worden bij het verbeteren/optimaliseren van de verwijdering van biologische/organische vervuilingspotentie van verschillende voorzuiveringsprocessen. De operationele omstandigheden werkzaam in mediafiltratie (zoals contacttijd, terugspoelprotocol, toevoeging van coagulatie enz.) waren van cruciaal belang bij het minimaliseren van biologische/organische vervuiling van SWRO membranen. Veelbelovende indicaties werden waargenomen bij het toepassen van de BGP methode als een biologische vervuilingsindicator in SWRO systemen.

Tot slot zijn de verdere ontwikkeling en toepassingen van de nieuwe ATP-gebaseerde BGP methode nog nodig in zeewater om de LOD te verlagen en om een drempelwaarde van BGP in SWRO voedingswater vast te stellen. Aangezien het BGP-protocol erg gecompliceerd is, zou het ontwikkelen van een online BGP-analysator erg gunstig kunnen zijn. Hiermee kan tevens de toepassingen van de BGP voor verschillende watertypen (zoet water, ballastwater enz.) gestimuleerd worden.

CONTENTS

1

INTRODUCTION AND THESIS OUTLINE

1.1 FRESHWATER AVAILABILITY

Water is essential for sustainable growth and maintaining healthy ecosystems. Unfortunately, water scarcity is among the main problems of the twenty first century experienced by many societies and countries, making it a worldwide problem. Throughout the years, shortage of water in the world has been increasing due to climate change and rapidly increasing population, which in turns requires more water for domestic, agricultural and industrial use. Figure 1.1 shows the available renewable freshwater in the world in 2013. It can be clearly seen that water scarcity is already an issue in many countries in North Africa, and the Arabian Peninsula, where the available freshwater is less than 1000 m^3/person/year. Almost 75 % of the Arab population live under the water scarcity level, and nearly 50 % lives under extreme water scarcity (UN report, 2015). In addition, many other countries in Asia, Africa and Europe including but not limited to India, Pakistan, South Africa, Zimbabwe, Ethiopia, Somalia, Poland, and Czech Republic are water stressed.

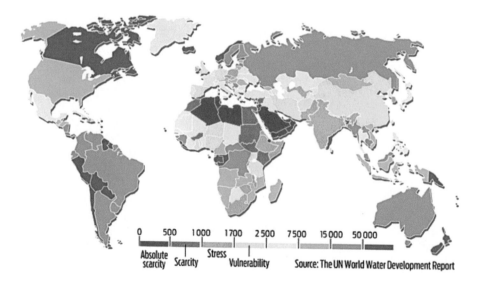

Figure 1.1: The available renewable freshwater in the world in 2013 in m^3 per person per year. (The UN world water development report, 2013).

To mitigate water shortages, many approaches have been used in the last few decades including water saving measures, reducing water loss in public water supply networks, wastewater reuse, rainwater harvesting, water transport, and desalination of brackish (ground) water and seawater.

1.2 SEAWATER DESALINATION

Seawater desalination is a promising solution to handle the challenge of meeting the water demand of an ever-increasing population. Seawater desalination is a water treatment process that separates salts from saline water to produce potable water. Seawater desalination techniques are mainly classified into two types: (1) processes based on physical change in the state of the water or distillate through evaporation, and (2) processes using a membrane that employ the concept of filtration (Al-Karaghouli et al. 2009).

Membrane technology provides a physical barrier that can effectively remove solids, bacteria, viruses, and inorganic compounds. Membranes are applied for surface water and wastewater treatment, desalination of seawater and brackish water and wastewater reuse. Reverse Osmosis (RO) is currently the most commonly used membrane technology for desalination (Aintablian 2017, DesalData 2018). Desalination capacity has been increasing over the past 50 years for all water sources (Brackish water, fresh water, seawater, wastewater, etc.) (DesalData 2018). The global desalination capacity continues to grow with 4.5 % per year (Figure 1.2).

Nowadays, more than 95 million cubic meters of all types of water is desalinated every day. Based on the DesalData desalination database of installed capacities in 2018, membrane desalination process (reverse osmosis/RO) is the leader in the market with 70 % (65 Mm^3/d) of total capacity in 2018 (Figure 1.2). The thermal process including multi-stage flash, multiple-effect desalination and vapour compression follows with 23 % (25 Mm^3/d) of the total capacity in 2018. The remaining capacity is divided between Electro-dialysis (ED) and other newer-concept systems.

3

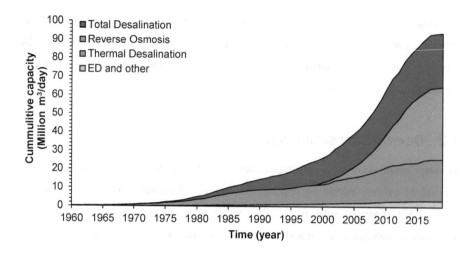

Figure 1.2: Growth of the desalination capacity (including seawater, brackish water, river water, wastewater, brine, and pure water desalination processes (DesalData 2018).

Approximately, 60 Million cubic meter per day is currently produced using seawater as a water source. Two-thirds of the desalinated seawater (40 Mm³/d) is produced using seawater reverse osmosis (SWRO) membrane in 4,300 SWRO desalination plants in all over the world (Figure 1.3), whereas the other one-third (20 Mm³/d) is produced using thermal process in 1,800 desalination plants.

The distribution of the SWRO desalination plants per geographical area based on the number and capacity of installed SWRO desalination plants is presented in Figure 1.4a and 1.4b, respectively. The highest installed desalination plants is in the Middle East and North Africa with 1,850 SWRO plants (42 %), which is equivalent to the installed capacity of the SWRO plants in North America (14 %), Eastern Europe (15 %), and East Asia (13 %) (Figure 1.4a). The distribution based on the capacity of SWRO desalination plants is more or less similar to its distribution based on the number of installed SWRO desalination plants per geographical area. The highest capacity of SWRO is in the Middle East and North Africa (45 %, 18.5 Mm³/d) (Figure 1.4.b).

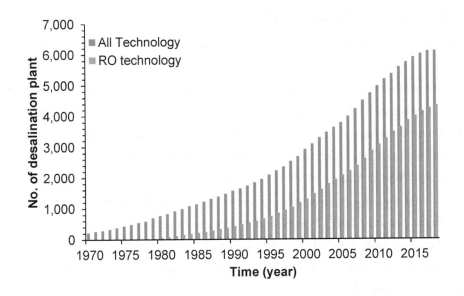

Figure 1.3: Growth of the installed seawater desalination plants (DesalData 2018).

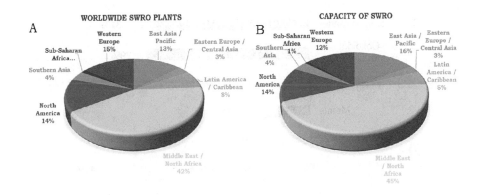

Figure 1.4: World seawater desalination per Geographical area based on (a) number of installed SWRO desalination plants and (b) capacity of the SWRO plants (DesalData 2018).

1.3 FOULING OF REVERSE OSMOSIS

Although RO is widely applied, membrane fouling is the most serious operational problem that face RO membrane systems. Membrane fouling is mainly caused by the deposition of material onto the membrane causing an increase in feed channel pressure drop or permeability decline, and increase in salt passage (Al-Ahmad et al. 2000). In membrane filtration application, fouling can be categorized according to the type of foulant: biological and organic fouling, particulate fouling (colloidal matter), and scaling (inorganic) (Abd El Aleem et al. 1998). One or more types of fouling can occur depending on the feed water quality, operation conditions of the SWRO membrane and type of membrane (She et al. 2016). Pena N. et al. (2013) performed autopsies on 600 damaged membrane elements from all over the world and found that 31 % of the damaged membranes were due to biofouling and around 22 % of membranes were due to scaling as shown in Figure 1.5.

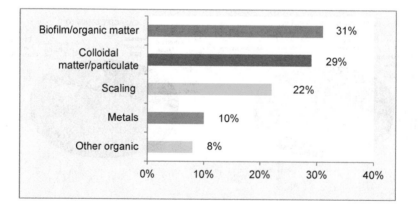

Figure 1.5: Main types of foulants detected on membrane autopsies of 600 RO modules (Pena N. et al. 2013).

Biofouling occurs when microorganisms grow in the membrane system by utilizing the biodegradable substances from the water phase and converting them into metabolic products and biomass. Microorganisms can multiply to form a thick layer of slime called a biofilm (Flemming 2011). Biofouling starts when biofilm on RO membrane leads to operational problems. Particulate fouling due to the deposition of colloidal and suspended matter is controlled by conventional treatment (coagulation, flocculation, and media filtration) process and ultrafiltration as a pre-treatment. Scaling of sparingly soluble salts (such as of $CaCO_3$, $CaSO_4$, $BaSO_4$ and SiO_2) occurs when the concentration of the ions in the water exceeds the saturation concentration. Scaling can be controlled to a certain degree by addition of acid (for only some scales; i.e. $CaCO_3$) and/or antiscalant.

1.4 MONITORING OF BIOFOULING

To date, there is no standard method to monitor biological/organic fouling in RO membrane system. However, several approaches are used to monitor biofilm development on RO membrane surfaces and biofouling potential in RO feed water.

The most commonly applied method to monitor biofouling in full-scale RO plants is the use of system performance data such as the development of pressure drop, permeability, and salt passage. However, it cannot be used to mitigate biofouling as biofouling has already occurred when observing any changes in the RO performance RO membrane.

Another practice to monitor biofouling is the direct detection of biofilm formation on RO membranes using online sensors such as electrical potential measurements (Sung et al. 2003), biosensors (Lee and Kim 2011), ultrasonic time-domain reflectometry (Kujundzic et al. 2007) and on-line fluorimeter (Ho et al. 2004). However, the major challenges on these methods are the fouling on these sensors over a long period of use, the need for frequent calibration of the sensors, the high pressure applied in RO system and the use of spiral wound modules (Nguyen et al. 2012a).

Additionally, some tools were proposed to monitor biofilm formation on RO membrane surfaces, which they are connected in parallel to the RO membrane unit such as the membrane fouling simulator (MFS) (Vrouwenvelder et al. 2006) and modified biofilm formation rate monitor (mBFR) (Ito et al. 2013). In these tools, biofilm formation is frequently monitored on a flat sheet membrane or Teflon/glass rings, in which feed flow is controlled to be close/similar to the feed flow in RO system. However, these methods

7

require a lot of time to detect the formation of a biofilm (the same time needed for biofilm formation of full-scale plant).

Another approach to monitor biofouling is to measure the biological/organic fouling potential through the pre-treatment and in the feed water of RO membrane systems. This approach is attractive because it can be used as an early warning system allowing adjustment of the operational conditions of the pre-treatment processes to meet the required quality in RO feed water and consequently better control of biofouling in RO systems. Such biological/organic fouling potential methods are assimilable organic carbon (AOC), biodegradable dissolved organic carbon (BDOC), bacterial growth potential (BGP), Liquid Chromatography – Organic Carbon Detection (LC-OCD), Transparent Exopolymer Particles (TEP).

1.5 CONTROLLING OF BIOFOULING

Even though membrane biofouling is an unavoidable phenomenon during membrane filtration, it can be minimized/delayed by optimizing the pre-treatment to significantly remove bacteria and nutrients/organics from RO feed water. In case of RO biofouling, RO performance is restored by membrane cleaning.

1.5.1 Pre-treatment of RO feed water

For successful operation and control of RO, the raw seawater needs to be pre-treated before it passes through the RO membrane system (Dietz and Kulinkina 2009). The key objective of the pre-treatment is (i) to improve the seawater quality of the RO feed water and (ii) to increase the efficiency and life expectancy of the membrane elements by minimizing fouling, scaling and degradation of the membrane.

Pre-treatment refers to the various physical and chemical water treatment processes normally including the use of sand filters and cartridge filters (Figure 1.6), when required (Mallick 2015). It may also include chemical treatment if scaling or fouling of the RO membranes is anticipated. Almost all SWRO desalination plants require pre-treatment processes. The source and quality of raw seawater plays a significant role in the level and type of pre-treatment required. Pre-treatment can be a very significant part of the overall plant infrastructure, especially for the poor quality raw seawater.

Nutrient removal from feed water is an approach to control biofouling. Nutrients in seawater results in substantial bacterial growth. Therefore, it is essential to limit nutrient concentration in order to prevent their logarithmic growth. Carbon is the major nutrient required by marine microorganisms. Total organic carbon (TOC) was measured to evaluate carbon availability in seawater. However, not all the TOC concentration is utilizable by bacteria. Figure 1.7 shows a schematic overview of the different organic carbon fractions in water. Only the bio-available fraction is used by bacteria for growth. The bio-available carbon fraction is measured as BDOC and/or AOC (Volk 2001). It has been reported that organic matter, more specifically AOC is mostly utilized by bacteria and is related to microbial growth and biofouling of membrane systems (Werner and Hambsch 1986, Weinrich et al. 2011).

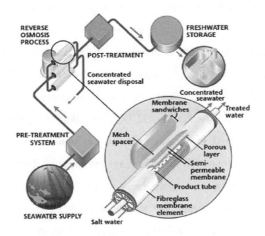

Figure 1.6: Stages of an RO membrane system (Kabarty 2016).

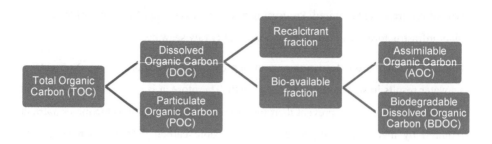

Figure 1.7: Schematic overview of the different organic carbon fractions in drinking water (Hammes 2008).

Several fractions of organic carbon have been studied over the years. Visvanathan et al. (2003) used dissolved organic carbon (DOC) and reported that using membrane bioreactor as pre-treatment for seawater RO can remove 78 % of DOC. In addition, Jeong et al. (2013a) studied the removal biopolymers and humics in a submerged membrane system coupled with PAC adsorption in saline water, and they reported a removal of biopolymers and humics in the range of (94 – 97 %) and (71 – 76 %), respectively. Weinrich et al. (2011) used AOC to investigate the organic removal through the RO pre-treatment in the Tampa Pay desalination plant and they found 65 % removal of AOC through the sand filtration. Van der Kooij et al. (2003)reported that the maximum AOC concentration in drinking water should be lower than 10 µg/L in order to prevent excessive bacterial growth in the water distribution network. Hijnen et al. (2009) found that the rate of biofouling on RO membrane systems depended on the AOC concentration present in the RO feed water, in which the AOC threshold concentration was 1 µg-C/L to avoid biofouling in fresh water RO membranes. However, this very low threshold concentration is very difficult to achieve through the pre-treatment. Weinrich et al. (2015) reported a preliminary AOC threshold concentration of 50 µg-C/L based on pilot testing.

1.5.2 Chemical Cleaning

In case of RO biofouling, membrane cleaning is performed to control biofouling. Membrane cleaning is usually carried out when there is significant drop of differential pressure drop and/or permeability (Vrouwenvelder and Van Loosdrecht 2009). The concept of cleaning is to remove and/or kill the accumulated biomass from the membrane surface, so the initial permeability can be recovered. Cleaning can be performed both as physical cleaning (flushing) and chemical cleaning (Nguyen et al. 2012b).

Generally, physical cleaning is applied prior to chemical cleaning (Cornelissen et al. 2009). Physical cleaning includes air/water flushing cleaning. Physical cleaning uses mechanical pressure which removes mostly non-adhesive fouling. Cornelissen et al. (2009) stated that air/water cleaning is an efficient way to control biofouling in spiral wound membranes and by applying air/water cleaning, biomass concentration was reduced by 83 %.

The second step of membrane cleaning is chemical cleaning which is usually cleaning in place (CIP) treatment. Several types of chemical cleaning agents (e.g. alkaline, acids, biocides, detergents, enzymes etc.) are recommended by membrane manufacturers (Wibisono et al. 2015). Chemical cleaning is efficient in killing or inactivating microorganism but it does not remove the accumulated biomass (Flemming 1997, Vrouwenvelder et al. 1998). As a consequence, the remaining inactivated biomass is consumed as food by other surviving bacteria causing rapid regrowth of bacteria in cleaned membranes (Vrouwenvelder et al. 2010). Thus, membrane cleaning both in terms of physical and chemical cleaning can partially reduce the biofouling for a short periods, but cannot guarantee biofouling control.

Chlorine is very efficient in killing bacteria, but it is limited in SWRO membrane applications because free chlorine can damage the membranes and lower salt rejection. Another drawback of using chlorine is that it breaks down the organic and humic substance to AOC which leads to rapid growth of biofilm, hence rapid increase of feed channel pressure drop. Mono-chloramine is also used to control biofouling in some treatment plants. Vrouwenvelder and Van Loosdrecht (2009) reported that a limited number of treatment plants are able to successfully control biofouling to some extent by dosing mono-chloramine. Nevertheless, use of mono-chloramine can form N-nitrosodimethylamine (NDMA), which is human carcinogenic substance. Also, water containing mono-chloramine can cause membrane damage in the presence of iron and manganese which catalyse the oxidation of membranes (Vrouwenvelder et al. 2010).

1.6 BACTERIAL GROWTH METHODS

Biofouling is difficult to control as it depends on many factors and it occurs undoubtedly due to the presence of microorganisms and nutrients in feed water. Monitoring bacterial growth potential is likely to be a good approach for biofouling control, but many different methods currently exist.

1.6.1 Assimilable organic carbon (AOC)

Assimilable organic carbon is the fraction of dissolved organic carbon, which is utilized by heterotrophic bacteria for their growth. The AOC concept and its bioassay was initially introduced by Van der Kooij et al. (1982). AOC contains many low molecular weight organic molecules namely sugar, organic acids and amino acids (Hammes and Egli 2005). It comprises a very small fraction (0.1 – 10 %) of total organic carbon present in the water and only a few micrograms of AOC can lead to significant bacterial growth and biofouling problems (Van der Kooij 1992). Hijnen et al. (2009b) reported that the AOC threshold concentration in feed water is around 1 µg/L for biofouling in fresh water reverse osmosis membranes which is very low and very difficult to achieve through pre-treatment. In addition, they found that the rate of biofouling is depended on the acetate-C concentration. Weinrich et al. (2009), reported that 10 µg/L of AOC in water results in significant heterotrophic bacterial growth, while Van der Kooij (1992), showed that there was no significant growth of bacteria in non-chlorinated water with AOC concentrations lower than 10 µg/L in drinking water systems. It can be seen that AOC is a critical parameter for drinking water treatment and a very low concentration of AOC can result in problems such as growth in distribution systems, biofilm formation and biofouling in reverse osmosis applications. Therefore, AOC measurements in seawater can also be used as a monitoring tool for bacterial growth and biofouling in RO systems.

The principle of the method is to inoculate a certain type of microorganism in a water sample. The sample is incubated at a specific temperature and the bacterial growth is monitored until it reaches the stationary phase. The max growth of bacteria is proportional to the AOC concentration in water and therefore it is converted to carbon concentration based on the bacterial yield. More than 15 different methods for AOC have been developed for freshwater over the past 20 years (Hammes et al. 2010a). Table 1.1 shows some of the methods for AOC measurement in freshwater. The differences between these methods is the enumeration method used to monitor bacterial growth (plating, turbidity, flow cytometry (FCM), adenosine triphosphate (ATP), etc) and the conversion of the measured data to AOC (growth/biomass).

AOC measurement is based on bacterial growth until the stationary phase is reached, which makes it time consuming, complex and laborious. For instance, several studies have been conducted to optimize the incubation period for AOC tests, which might take a few days (12-14 days for the conventional AOC method). In these studies, the inoculum selection, incubation time and the bacterial growth enumeration have been investigated (Wang et al. 2014). The conventional AOC method was developed by Van der Kooij et al. (1982), in which two strains of bacteria: *Pseudomonas fluorescens* (P17) and *Spirillumsp NOX* (NOX) are used as inoculum and bacterial growth is measured by heterotrophic plate counting. Initially, Van der Kooij used *Pseudomonas fluorescens* (P17) as inoculum for AOC bioassay. However, the use of one strain as inoculum may not represent the utilization of AOC in water. For this reason, another strain (NOX) was added to P17 as inoculum to broaden the AOC utilization. 500 CFU/mL of bacteria (P17 + NOX) are inoculated and then incubated at 15 °C for 9 days. Bacterial growth is monitored by heterotrophic plate counts and the net bacterial growth for both strains P17 and NOX is then converted to AOC concentrations as acetate-C equivalents based on bacterial yield (4.6×10^6 for P17 and 1.2×10^7 for NOX).

Table 1.1: Developed method for AOC measurement in fresh water (Wang et al. 2014).

Author/year	Inoculum	Bacterial growth measurement	Test time (days)
Van der Kooij et al. (1982)	P17	Plate counting	12-14
Werner and Hambsch (1986)	Indigenous microorganism	turbidity measurement	2-4
Kemmy et al. (1989)	Four bacteria	Plate counting	5
Van der Kooij (1992)	P17, NOX	Plate counting	12-14
Kaplan et al. (1993)	P17, NOX	Plate counting	5-9
LeChevallier et al. (1993)	P17, NOX	ATP	3
Escobar and Randall (2000)	P17, NOX	Plate counting	9
Hammes and Egli (2005)	Indigenous microorganism	Total cell count with fluorescence staining using flow cytometry	5
Weinrich et al. (2009)	Bioluminescent	Bioluminescence	2-3

AOC methods in seawater:

Most AOC methods were developed for fresh water and they have been applied for conventional water treatment technology and biological stability assessment of drinking water and reclaimed waste water treatment (Wang et al. 2014). Recently, Weinrich et al. (2011) established an AOC method in saline water using *Vibrio harveyi* bacteria as inoculum. Also, Jeong et al. (2013a) developed a method to measure AOC concentration in seawater using a marine bacterium (*Vibrio fischeri*). In these two methods, bacterial growth is measured by bioluminescence and it is claimed by the author that the AOC measurements can be obtained within 1 - 8 hour. The method seems promising since it is rapid, but the complete utilization of AOC by a specific consortium of bacteria within such a short time is doubtful.

Table 1.2: Methods developed for AOC measurement in seawater.

Author/year	Target	Culture	Incubation time	Cell counts	Substrate
Weinrich et al. (2011)	Saline water	*Vibrio harveyi*	<1 d	Luminescence	Acetate
Jeong et al. (2013a)	seawater	*Vibrio fischeri*	<1 h	Luminescence	Glucose

1.6.2 Biodegradable organic carbon (BDOC)

BDOC is the total amount of organic matter that is biodegraded or assimilated by an inoculum of suspended or fixed bacteria over a period of time (Escobar and Randall 2001). The main difference between AOC and BDOC assays is that the BDOC assay assess the concentration of DOC removed through usually biofilm related microbial growth. The AOC assay usually assess the amount of cells produced through utilization of bio-available carbon (Hammes 2008). Escobar and Randall (2001) proposed that both AOC and BDOC be used as complementary measurements of bacterial regrowth to avoid potential over-estimation and under-estimation of biological stability.

1.6.3 Biomass production potential (BPP)

BPP is a modification of the AOC test described by Stanfield and Jago (1987) in which the water samples are inoculated with 1 mL of river water and incubated, without any treatment, at 25°C in the dark without shaking (Van der Kooij and Van der Wielen 2013). The biomass was monitored by measuring ATP concentration daily in the water during an incubation period of 14 days or longer. This test has been developed to evaluate the bacterial growth in drinking water networks. However, the application of this test in seawater may be limited since seawater may include a significant concentration of algal ATP. In the BPP method, the maximum microbial growth was observed within one week of incubation and the cumulative biomass production after 14 or 28 days of incubation are reported, without conversion to carbon concentration. Van der Kooij et al. (2017) found a strong significant correlation ($R^2 = 0.99$) between the maximum growth observed within one week and the cumulative biomass production after 14 days which may suggest the use of maximum growth within one week for faster biomass indication.

1.7 BACTERIAL ENUMERATION METHODS

Various methods are used to monitor microbial growth in water. The following methods are commonly used to determine the concentration of biomass in water or on membrane surface; heterotrophic plate counts (HPC), total direct cell counts (TDC), Total cell count (TCC) by FCM and ATP.

1.7.1 Heterotrophic Plate Count

HPC is a method to measure the heterotrophic microorganism population. Heterotrophs are organisms including bacteria, yeasts and moulds. These organisms require organic carbon for growth as an external source. HPC measures in colony forming unit per square centimetre (CFU/cm^2) by distributing 0.05 ml of water on nutrient poor R2A medium then counting the number of colonies manually after a few days of incubation at 25 °C. HPC can detect all microbial cells or all pathogenic bacteria. It is considered as a limited test or secondary indicator of TDC due to limited range and information that it can provide (Greenspan 2011).

1.7.2 Total Direct Cell counts

TDC is a rapid assessment method of all microbial cells (dead and active) in water samples. TDC unit is the number of counted cells per square centimetre (cells/cm^2) which can be determined microscopically using a fluorescent dye (acridine orange) (Hobbie et al. 1977). Although TDC is a simple method, it is time-consuming.

1.7.3 Cell concentration using flow cytometer

FCM is a technique of quantitative single cell analysis by counting and examining microscopic particles, such as cells and chromosomes. Fluorescent staining should be used before measuring. Berney et al. (2008) used staining with the nucleic acid-binding SYTO dyes and propidium iodide (PI) to distinguish between intact cells and damaged cells. FCM is widely used to study microbial growth in drinking water environments because it is a rapid and accurate. Hammes et al. (2008) developed a method to accurate quantify microbial cells at concentrations lower than 1,000 cells/mL. They found no correlation between Total Cell Concentration TCC-FCM and HPC while a good correlation was found between TCC-FCM and ATP especially between intact cells and microbial ATP.

1.8 ADENOSINE TRIPHOSPHATE (ATP)

ATP is a substance present in all living cells (including bacteria) that provides energy for many metabolic processes. In particular, it uses as a coenzyme in living cells and it is often called the "molecular unit of currency" of intracellular energy transfer (Knowles 1980). ATP consists of adenosine and three phosphate groups. Adenosine is composed of an adenine ring and a ribose sugar. The critical part of ATP is the phosphorous part - the triphosphate. Therefore, ATP has a lot of potential energy. ATP converts to adenosine diphosphate (ADP) if one of the three phosphates is broken down. This conversion is an extremely crucial reaction due to the energy released by the reaction. The reaction and realized energy of one mole ATP is shown in eq (1-1).

$$ATP + H_2O \rightarrow ADP + P_i \ \Delta G^\circ = -30.5 \ kJ/mol \ (-7.3 \ kcal/mol) \hspace{2cm} (1\text{-}1)$$

ATP is present in all bacterial cells and it is used as a parameter to quantify the amount of active biomass in water and also on surfaces (Velten et al. 2007). Many studies on the behaviour and presence of bacteria in the environment have confirmed the concept of using ATP as a measure of active bacteria (Van der Kooij et al. 2003). Today, ATP is widely applied to test the efficiency of treatment process at water treatment plants and to assess biological stability and after growth of drinking water in distribution systems (Vang et al. 2014).

ATP can be measured by serval methods such as (Khlyntseva et al. 2009): (i) Chromatographic methods, (ii) Fluorescence methods, (iii) Bioluminescence methods, and (iv) Sensors with immobilised luciferase. Among all, bioluminescence is the most common and attractive technique since it is the most rapid, sensitive and reproducible assay for measuring ATP content in water (Van der Kooij et al. 2003). ATP bioluminescence has been used to determining levels of ATP in many different cell types. The bioluminescence method involving the Luciferase enzyme is a multistep process which mainly requires Luciferin substrate, oxygen (O^2), magnesium (Mg^{2+}) and ATP.

17

Luciferase converts, in the presence of ATP and magnesium firefly D-luciferin into the corresponding enzyme-bound luciferil adenylate which converts to oxyluciferin in the presence of oxygen. This process occurs according to the following chemical equations:

$$D - luciferin + luciferase + ATP \xrightarrow{\text{Mg}} \text{Luciferil adenylate complex} + PPi \qquad (1\text{-}2)$$

$$\text{Luciferil adenylate complex} \xrightarrow{O_2} \text{Oxyluciferin} + AMP + CO_2 + light \qquad (1\text{-}3)$$

The light emission results from a rapid loss of energy of the oxyluciferine molecule. The light emission is in the range between 500 to 700 nm wavelengths. Under optimum conditions, light intensity is linearly related to the ATP concentration. Cellular ATP can be measured by direct lysis of the cells with a suitable detergent.

1.9 MOTIVATION OF THE STUDY

Controlling biofouling in SWRO membranes at an early stage is key to the successful and cost-effective operation of membrane-based desalination plants. Biofouling of SWRO membranes occurs due to the accumulation of biofilm to such an extent that causes operational problems in SWRO membrane systems. To date, no single parameter is available that can predict biofouling in membrane-based desalination systems. Biomass quantification is only used as a first indication of biofouling potential (Vrouwenvelder et al. 1998, Vrouwenvelder et al. 2008) as bacteria are always present in RO feed water even after ultrafiltration pre-treatment (Ferrer et al. 2015). Moreover, biofilm formation on RO is inevitable as long as the feed water supports significant bacterial growth due to the presence of dissolved nutrients. Hence bacterial growth potential of RO feed water has gained more attention than the removal of bacteria (LeChevallier 1990, Jeong et al. 2013b).

Several bacterial growth potential methods have been developed in freshwater such as AOC (Van der Kooij et al. 1982), BDOC (Joret and Lévi 1986) and BPP (Van der Kooij and Van der Wielen 2013). The relationship between these methods and biofouling development in full-scale plants has not yet been determined. In freshwater, Hijnen et al.(2009) reported that 1 µg/L of AOC (as acetate) added to MFS feed water led to a significant pressure drop within 3 months. In seawater, an attempt has been made by Weinrich et al. (2016), in which more biofilm formation was detected on a flat sheet RO membrane when AOC concentration of the feed water increased from 30 to 1,000 µg/L. In addition, differential pressure increased from 3.5 to 6.2 bar during 9 days of pilot testing when the median AOC was 50 µg/L.

In seawater, two AOC methods have been developed recently to measure the growth potential in the pre-treatment and SWRO feed water by Weinrich et al. (2011) and Jeong et al. (2013b), using a single strain of bacteria (*Vibrio fischeri* and *Vibrio harveyi*, respectively). The use of a single bacterial strain allows normalization of the yield based on a carbon source, enabling conversion of bacterial growth to a carbon concentration. However, this method may not reflect the carbon utilization of indigenous microorganisms in seawater, and thus it may underestimate the nutrients concentration of seawater. Developing a new method making use of an indigenous microbial consortium may provide more accurate and representative bacterial growth for seawater. However, a fast and accurate bacterial enumeration method in seawater is also need. The only available growth potential method in seawater that uses an indigenous microbial consortium is the BRP method developed by Dixon (Dixon et al. 2012), which is based on turbidity.

The bacterial enumeration method used to monitor the growth potential depends on the bacterial culture(s) used in the BGP test. Weinrich et al. (2011) and Jeong et al. (2013b) used bioluminescence to monitor the bacterial growth as both methods employed luminescent bacteria (*Vibrio fischeri* and *Vibrio harveyi*, respectively). Since, conventional enumeration methods (i.e. heterotrophic plate counting, total direct cell count) are laborious, time consuming and limited to a small percentage of the overall bacterial count (Jannasch and Jones 1959). Recently, new alternative methods that are culture-independent have been developed, such as flow cytometry (FCM) and ATP. FCM is fast, accurate, and can differentiate between intact and dead cells, nevertheless, it is recommended as a relative method because of the use of a manual gate to distinguish bacterial cells from other microorganisms, particles and the background of the machine (Montes et al. 2006, Thompson and van den Engh 2016). ATP is directly related to the activity of biomass (Holm-Hansen and Booth 1966, Karl 1980, Hijnen et al. 2009a). ATP has been widely used for freshwater applications. However, there is no method to quantify ATP in seawater because the luciferase-luciferine reaction is hampered by the presence of salts. Therefore, a new method for microbial ATP measurement in seawater is required. Using microbial ATP (as an enumeration method of the indigenous microbial consortium) to measure bacterial growth potential in seawater may provide more accurate and representative bacterial growth.

1.10 RESEARCH OBJECTIVES

The main goal of this research is to develop a method to measure bacterial growth potential (BGP) in seawater based on microbial Adenosine Tri-phosphate (ATP), making use of an indigenous microbial consortium. The method is intended to be used to monitor the removal of biological/organic fouling potential during SWRO pre-treatment. The specific objectives are:

1. To develop a method to measure low levels of ATP in seawater without interference from the seawater matrix.
2. To develop a microbial ATP-based method to measure bacterial growth potential in seawater using an indigenous microbial consortium.
3. To apply the developed methods to monitor the pre-treatment systems in full-scale SWRO desalination plants.
4. To investigate if any correlation exists between bacterial growth potential of SWRO feed water and SWRO membrane performance.

1.11 THESIS FRAMEWORK

This thesis is made up of seven chapters including six articles presenting the results and findings of the different segments of the research.

- After this introductory chapter, **Chapter 2** describes the development of a direct method of measuring total, free and microbial ATP in seawater.

- **Chapter 3 describes the** development of a filtration-based method to measure ATP in seawater, SWRO feed water and permeate. .

- **Chapter 4** presents the development of a method to measure BGP in seawater based on microbial ATP and using an indigenous microbial inoculum. This chapter also shows the relationship between BGP in SWRO feed water and the cleaning frequency of SWRO membrane systems.

- **Chapter 5** assesses the pre-treatment of a full-scale SWRO desalination plant in the Middle East in terms of particulate, biological and organic fouling potential removal. Moreover, this chapter presents the correlation between BGP in SWRO feed water and membrane performance.

- **Chapter 6** evaluates the removal of fouling potential in seawater dual media filters and dissolved air flotation used as pre-treatment in SWRO. The evaluation is performed using particulate, biological and organic fouling potential methods in seawater.

- Finally, **Chapter 7** provides a summary of the main conclusions, outlook and recommendations for further research.

2

DIRECT MEASUREMENT OF ATP IN SEAWATER

The use of adenosine triphosphate (ATP) to monitor bacterial growth potential of seawater is currently not possible, as ATP cannot be accurately measured at low concentration in seawater using commercially available luciferase-based ATP detection. The limitation is due to interference of salt with the luciferin-luciferase reaction, which inhibits light production. This research demonstrates that new reagents developed for (i) ATP extraction from microbial cells and (ii) ATP detection in seawater are able to reliably detect microbial ATP as low as 0.3 ng-ATP/L in seawater. The luminescence signal of the new detection reagent is significantly higher (> 20 times) than the luminescence signal of the freshwater reagent, when applied in seawater. ATP can now be used to monitor bacterial growth potential through pre-treatment trains of seawater reverse osmosis (SWRO) plants as the level of detection is significantly lower than required to prevent biological fouling in reverse osmosis membrane systems.

The new reagents have been used to monitor microbial ATP in coastal North Sea water. Moreover, microbial ATP has been applied to monitor the bacterial growth potential (using indigenous bacteria) through the pre-treatment train of an SWRO desalination plant. A significant reduction (> 55 %) of the bacterial growth potential was found through the dual media filtration with 4.5 mg-Fe^{3+}/L coagulant. Overall, the new reagents can detect low microbial ATP concentrations in seawater and can be used to monitor bacterial growth potential in seawater desalination plants.

Keywords: adenosine triphosphate, seawater, bacterial growth potential, reverse osmosis, desalination, biofouling.

This chapter has been published as **Almotasembellah Abushaban**, M. Nasir Mangal, Sergio G. Salinas-Rodriguez, Chidiebere Nnebuo, Subhanjan Mondal, Said A. Goueli, Jan C. Schippers, Maria D. Kennedy. Direct measurement of ATP in seawater and application of ATP to monitor bacterial growth potential in SWRO pre-treatment systems. *Desalination and Water Treatment* (2017) 99 p 91–101.

2.1 INTRODUCTION

Controlling biological fouling in SWRO membranes at an early stage is key to the successful and cost-effective operation of membrane-based desalination plants. Biofouling of SWRO membranes occurs due to the accumulation of biofilm on the membrane surface, or accumulation across the spacer-filled membrane feed channels to such an extent that the operational problem threshold is exceeded, typically a 15 % reduction in initial performance. Operational issues may include an increase in pressure drop across the elements, an increase in salt passage and membrane degradation. To mitigate most of these problems, plant operators clean the membranes as frequently as the biofouling threshold is exceeded. The cleaning in place (CIP) is performed by soaking and flushing the membrane channels with various chemicals in an attempt to remove the biofilm. The frequency of CIP is site specific and varies from time to time, depending mainly on the biofouling potential of the seawater source, operational conditions and the effectiveness of the pre-treatment processes in removing readily available nutrients.

To date, no single parameter is available that can predict biofouling in membrane-based desalination systems. Biomass quantification is only used as a first indication of biofouling potential (Vrouwenvelder et al. 1998, Vrouwenvelder et al. 2008) as bacteria are always present in RO feedwater even after ultrafiltration pre-treatment (Ferrer et al. 2015). Moreover, biofilm formation in RO is inevitable if the feedwater supports significant bacterial growth due to the presence of dissolved nutrients. Hence, bacterial growth potential of RO feedwater has gained more attention than the removal of bacteria itself (LeChevallier 1990, Jeong et al. 2013b). Several methods directly linked to bacterial growth have been developed such as assimilable organic carbon (AOC) (Van der Kooij et al. 1982), biodegradable dissolved organic carbon (BDOC) (Joret and Lévi 1986) and biomass production potential (BPP) (Stanfield and Jago 1987, Van der Kooij and Van der Wielen 2013). Weinrich et al. (2016) detected more biofouling (using a flat sheet RO membrane) when the AOC concentration of the feed water increased from 30 to 1,000 µg/L. In addition, differential pressure increased from 3.5 to 6.2 bar during 9 days of pilot testing when the median AOC was 50 µg/L.

AOC measurements have been widely studied for potential applications involving freshwater employing heterotrophic plate counting (Van der Kooij et al. 1982, Kemmy et al. 1989, Kaplan et al. 1993, Escobar and Randall 2000), flow cytometry (Hammes and Egli 2005), ATP measurement (Stanfield and Jago 1987, LeChevallier et al. 1993), and bioluminescence (Haddix et al. 2004). Most of these AOC methods used a pure strain of bacteria as inoculum, and the first AOC method using indigenous bacteria in freshwater was developed by Stanfield and Jago (1987) and ATP was used for bacterial enumeration. Ross et al. (2013) reported higher (> 20 %) bacterial growth in freshwater when using indigenous bacteria compared with a pure strain. Similar AOC studies for seawater have lagged behind compared with freshwater. However, Weinrich et al. (2011) adapted the AOC-bioluminescence freshwater method for seawater by using a specific strain of bioluminescent marine bacteria, *Vibrio harveyi*. Jeong et al. (2013b) found a strong correlation between the number of another single bioluminescent strain, Vibrio fischeri and the bioluminescence signal (in the range of 10^3 to 10^5 CFU). Consequently, the bioluminescence of a single strain of bacteria has been increasingly adopted for AOC measurement in seawater. These two methods are fast (1 hour and 1-3 days, respectively) but use a pure strain of a single bacterium which may not reflect the carbon utilization of a natural bacterial community in seawater. It should be noted that these two methods cannot be applied with indigenous bacteria because not all naturally occurring bacteria show bioluminescence. Developing an AOC method using indigenous bacteria for seawater (similar to the Stanfield and Jago method for freshwater) may provide results with more predictive value in terms of biofouling in SWRO than the use of a pure bacterial strain.

Several methods can be used in seawater to monitor bacterial growth including heterotrophic plate count (HPC), total direct count by microscope (TDC), flow cytometry (FCM), and ATP. HPC is laborious, time consuming, and limited to the enumeration of cultivable bacteria (Staley and Konopka 1985, Liu et al. 2013b). TDC does not distinguish between active and inactive cells and is limited to samples that have high cell concentrations (> 10^7 cell/mL) (Postgate 1969). FCM is fast, accurate, and can differentiate between intact and dead cells, nevertheless, it is recommended as a relative method because of the use of a manual gate to distinguish the bacterial cells from other microorganisms, particles and the background of the machine (Montes et al. 2006, Thompson and van den Engh 2016).

ATP is known as the "energy currency" of cells (Knowles 1980, Webster et al. 1985) as it is present in all living cells and rapidly degrades when cells die (Karl 1980). Thus, ATP is directly related to the activity of biomass (Holm-Hansen and Booth 1966, Karl 1980, Hijnen et al. 2009a). ATP in a given water sample can be classified into two separate fractions: Microbial ATP and free ATP. Microbial ATP is present within the living cellular population in the sample. Free ATP is present outside the cell, which can be generated from the release of cellular ATP upon cell death. ATP has been used to assess microbial activity in drinking water, groundwater, biofilms in distribution networks and to monitor freshwater treatment processes (Magic-Knezev and Van der Kooij 2004, Vrouwenvelder et al. 2008, Hammes et al. 2010b). In freshwater RO systems, ATP has been applied as a biomass parameter: (i) to quantify biomass on membrane surfaces and diagnose biofouling (Vrouwenvelder et al. 1998, Vrouwenvelder et al. 2008), (ii) to measure biomass in the feed water (Veza et al. 2008), and (iii) as a biomass parameter in bacterial growth potential measurements (Stanfield and Jago 1987).

There are no commercially available ATP kits for seawater due to interference from salts. The high ionic strength of seawater has been demonstrated to cause substantial inhibition of the enzymatic ATP reaction, so that the emitted light signal interferes with the background luminescence (Amy et al. 2011, Van Slooten et al. 2015). Van der Kooij in Amy et al. (2011) suggested diluting seawater with demineralised water to an electrical conductivity of 4 mS/cm (2.5 mg/L) to avoid salts interference. However, diluting seawater also substantially lowers the biomass concentration (ATP) which in turn limits the use of the method to samples with high biomass concentrations. Moreover, bacterial cells may burst at low electrical conductivity due to the osmotic pressure shock, and consequently only total ATP can be determined. Due to lack of an ATP method for seawater, an attempt was made by Simon et al. (2013b) to use BacTiter-Glo reagent (freshwater regent) to measure ATP at the inlet and outlet of a lab scale biofilter for seawater with reported high ATP concentrations. LOD using the freshwater reagent kit (BacTiter-Glo) in seawater was investigated in our group to be 50 ng-ATP/L (2×10^5 cells/L). Van Slooten et al. (2015) developed a method based on filtration to quantify

ATP of large organisms (10–50 μm) present in ballast water. In this method, organisms are concentrated on a 10 μm filter. Thereafter, the filter is placed in a cuvette with sterile Milli-Q water (Millipore) to concentrate the organisms in a small volume of demineralized water. Limitations are that the method is time consuming and exposing the marine organisms to demineralized water may result in bacterial osmotic shock which underestimates the ATP concentration.

The objective of this chapter is to illustrate the applicability of new reagents (Water-Glo kit) developed by Promega for microbial lysis and ATP detection in seawater. The microbial ATP measurement in seawater is intended for monitoring of bacterial growth potential in the pre-treatment and feed of SWRO systems using indigenous bacteria.

To establish the target limit of detection necessary in seawater reverse osmosis systems, the lowest threshold AOC concentration to avoid biofouling in RO membranes was used. Hijnen et al. (2009a) reported 1 μg-acetate/L in RO feed water as the threshold value to avoid biofouling in freshwater RO membrane systems. AOC concentration was converted into cell concentration using the conversion factor (1 μg-acetate/L = 1×10^4 cells.mL^{-1}) reported by Hammes et al. (2010a). Assuming that this is applicable to seawater, the ideal method would allow the detection of ATP in seawater samples down to 2.5 ng-ATP/L (using a conversion factor of 1×10^4 cells/mL = 2.5 ng-ATP/L reported in this research - see Fig 2.3b).

The following aspects have been investigated and are described in this chapter:

1 Verifying the luminescence signal and stability of the new detection reagents in seawater.

2 Testing the efficiency of the new lysis and detection reagents in seawater.

3 Testing the effect of seawater pH and iron concentration on the luminescence signal.

4 Calibration curve and the limit of detection of the measurement.

5 Monitoring microbial ATP in raw seawater.

6 Measuring microbial ATP and bacterial growth potential along the pre-treatment train of an SWRO desalination plant using an indigenous bacterial consortium.

2.2 MATERIALS AND METHODS

2.1.1 Sample collection, transportation and storage

Coastal seawater samples were collected in Jacobahaven (Kamperland, The Netherlands) between January and December 2016. All samples were collected in sterile 500 mL glass sampling bottles with tight-fitting screw caps and transported for 120 km in a cooler at 5 °C for analysis. The characteristics of the tested seawater are shown in Table 2.1.

Table 2.1: Water quality of seawater from Kamperland (The Netherlands, North Sea)

Parameter	Value
pH	7.9 ± 0.1
TDS (g/L)	32.5 ± 0.8
Conductivity (mS/cm)	52.6 ± 1.2
Total bacterial count (cells/mL)*	$1.2 \pm 0.48 \times 10^6$
TOC (mg-C/L)	1.28 ± 0.85
UV_{254} (1/cm)	0.045 ± 0.009

*measured with FCM

2.1.2 Preparation of artificial seawater (ASW)

ASW was prepared using Milli-Q water and analytical-grade inorganic salts (Merck, USA) with ion concentrations similar to the average global seawater (Villacorte 2014) (23.67 g/L $NaCl$, 10.87 g/L $MgCl_2.6H_2O$, 4.0 g/L Na_2SO_4, 1.54 g/L $CaCl_2.2H_2O$, 0.74 g/L KCl, 0.21 g/L $NaHCO_3$, and 0.002 g/L Na_2CO_3). The pH, electrical conductivity, and total ATP of ASW was 8.0 ± 0.1, 52.6 ± 1.2, and < 0.05 ng-ATP/L respectively.

2.1.3 Measurement of microbial ATP in seawater

In this method, microbial ATP was extracted directly by adding ATP Water-Glo lysis reagent (ATP Water-Glo Kit, Promega Corp., USA) to the seawater sample. Both, total ATP and free ATP were measured separately to determine the microbial ATP (microbial

ATP = total ATP - free ATP). The manufacturer of the reagents recommends that the volume of seawater sample plus the volume of lysis reagent should be equal to or less than the same volume of water detection reagent.

To measure total ATP concentration, 100 µL of ATP Water-Glo lysis reagent (Promega Corp., USA) was added directly to 100 µL of the seawater sample in a 1.5 mL micro centrifuge tube (sterile Eppendorf tube, Sigma-Aldrich). The mixture was heated at 38 °C for 5 min. Following the manufacturer's recommendation, an aliquot of 200 µL of preheated ATP Water-Glo detection reagent (Water-Glo, Promega Crop., USA) was added to the mixture and then the luminescence was recorded using a luminometer (GloMax®-20/20, Promega Corp.). To measure free ATP concentration, the same procedure was followed, but, without the addition of the ATP Water-Glo lysis reagent. The measured luminescence signal was converted to ng-ATP/L using two different calibration lines; one for total ATP and the second one for free ATP. As the solution matrix is different in each case, a separate calibration line is needed. Calibration lines with ATP concentration ranging from 0 to 500 ng-ATP/L were prepared using ATP standard (100 nM, Promega Corp., USA) and autoclaved seawater. The free ATP concentration was subtracted from the total ATP concentration to get the microbial ATP concentration. All analyses were performed in triplicate.

2.1.4 Testing the efficiency of lysis and detection reagent

The lysis effectiveness of the reagent in raw seawater was studied in two parts. The first part compares the new Water-Glo lysis reagent with chlorine. The second part investigated the concentration of microbial cells that can be effectively lysed by Water-Glo lysis reagent.

The lysis efficiency of Water-Glo lysis reagent (Water-Glo kit, Promega Corp., USA) and free chlorine (8 mg Cl_2/L) were compared based on the microbial ATP concentration of a seawater sample collected from the North Sea (the Netherlands). The sample was filtered over a 0.1 µm filter (sterilized, Millipore) to accumulate the microorganisms on the membrane surface and thereafter the lysis reagent/solution was filtered through the same filter to extract microbial ATP from the accumulated microorganisms on the membrane surface. The extracted microbial ATP concentration in the filtrate was then measured according to the free ATP protocol described in section 2.2.3.

To prepare the free chlorine solution, sodium hypochlorite (3.5 % Cl_2) was diluted (2,500x) in ASW to obtain 8 mg-Cl_2/L of free chlorine. The sample was neutralized (5 min contact time) by adding 10 mM $Na_2S_2O_3$ with a 1 % (v/v) ratio (Nescerecka et al. 2016). The concentration of chlorine was selected based on the findings of Nescerecka et al. (2016) where it was observed that a range of free chlorine concentration (5.6-11.2 mg-Cl_2/L) could completely extract microbial ATP without oxidation of ATP molecules in freshwater.

For the second part of the lysis effectiveness study, marine microorganisms present in 1 L of raw seawater collected from North Sea was concentrated in 30 mL by filtering the seawater (1 L) through a 0.2 μm filter using a vacuum pump. The concentrated marine microorganisms were re-suspended in 30 mL of the same seawater sample. The intact cell concentration was then measured by flow cytometry and the concentrated sample was diluted in the filtered raw seawater to get a different concentration of microorganism (1.2×10^5 to 1.2×10^7 cells/mL).

To investigate whether sufficient Water-Glo detection reagent is present for the ATP reaction, microbial ATP of a seawater sample was measured for the recommended volume ratio of seawater sample to lysis reagent to detection reagent (100: 100: 200 μL) and compared with two different volume ratios (100: 100: 100 μL, 100: 200: 300 μL), including different volumes of detection reagent. As different total volumes and amount of reagents are used, calibration lines of free ATP and total ATP were established to determine the microbial ATP concentration.

2.1.5 Effect of pH and iron concentration on the luminescence

To study the effect of pH, the luminescence signal of ASW at different pHs, ranging from pH 7 to 8.5, was measured. The pH of ASW was adjusted with 0.03M HCl (37 %, ACROS organics) and 0.03M NaOH (J.T. Baker). The prepared ASW was filtered through 0.1 μm filter (sterilized, Millipore) to remove bacteria that might be introduced during the pH adjustment. Thereafter, the protocol of total ATP measurement was followed to measure the luminescence signal.

Similarly, to study the effect of iron, the luminescence signal of ASW in the presence of iron concentrations (0, 0.1, 0.3, 0.5, 1, 2, 3, 5, and 10 mg/L) was measured following the protocol of total ATP. To prepare different concentrations of iron, a stock solution with 2M $FeCl_3.6\ H_2O$ (Merck Millipore) was prepared using ASW.

2.1.6 Monitoring of ATP and bacterial growth in an SWRO plant

Microbial ATP and bacterial growth potential (based on microbial ATP) were monitored in an SWRO desalination plant in Australia. The RO pre-treatment processes include a drum screen, coagulation and flocculation, dual media filter (DMF), and cartridge filter. Four samples were collected in October 2016 (spring season) through the RO pre-treatment (Fig. 2.1); just before coagulation (raw seawater), after coagulation and flocculation, after DMF, and after cartridge filter.

For bacterial growth monitoring, the samples were pasteurized (for 30 min) and 15 mL was transferred into 30 mL AOC-free vials (heated in an oven furnace for 6 hours at 550 °C) in triplicate. In order to broaden the bacterial versatility, a natural consortium of a bacterial population from the same location (as the sample) was inoculated (~200 µL inoculum volume) with an initial bacterial cell density of 1×10^4 intact-cells/mL (measured by flow cytometry) in each vial. The samples were incubated at 30 °C. The bacterial growth of the seawater sample was monitored for 5 days using the ATP protocol described in Section 2.2.3.

2.1.7 Determination of limit of detection (LOD)

The LOD was determined for both total ATP and free ATP based on an average of 10 blanks plus three times the standard deviation of the blank (Taverniers et al. 2004). The LOD of the microbial ATP method was calculated using the combined procedure, which is the square root of the sum of the squares of free ATP and total ATP ($LOD = \sqrt{LOD\ of\ Free\ ATP^2 + LOD\ of\ Total\ ATP^2}$).

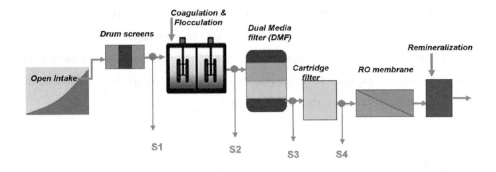

Figure 2.1: The treatment scheme of the tested SWRO desalination plant in Australia and the locations of the collected samples.

2.3 RESULTS AND DISCUSSION

2.3.1 Luminescence signal and stability of the new reagents

The luminescence signal and stability of ATP Water-Glo lysis and detection reagents were tested in seawater and compared to BacTiter-Glo (combined freshwater reagent). The ATP Water-Glo reagent showed higher luminescence signal (> 20x) compared to that obtained with BacTiter-Glo, when applied in seawater (Fig 2.2a). ATP Water-Glo reagent showed a good correlation between ATP concentration and luminescence signal with an R^2 of 0.99. It was also noted that the luminescence background of the ATP Water-Glo reagent is much lower (515 RLU) compared to that of BacTiter-Glo (2,263 RLU) in artificial sweater (35 g/L). The high luminescence signal and low background luminescence of the ATP Water-Glo reagents suggest that the new reagents provides more sensitivity than BacTiter-Glo (freshwater reagent) when used in seawater.

A thermostable firefly luciferase is used in formulating the ATP Water-Glo reagent. The ATP Water-Glo reagent is provided as a lyophilized substrate containing a mixture of luciferase and luciferin and a reconstitution buffer. Upon reconstitution, the stability of the liquid reagents was tested when stored at 4 °C and 23 °C and then compared with the stability of BacTiter-Glo at 23 °C. The ATP Water-Glo reagent retained over 90 % of its activity for 1 month at 4 °C and for 10 days at 23 °C, whereas the activity of BacTiter-

Glo dramatically decreased within the first day (Figure 2.2b). The stability of the luminescence signal was also tested, and was stable for 40 seconds after the addition of ATP Water-Glo reagent to the seawater sample (Fig 2.2c). These results demonstrate that the new reagents are suitable for application for seawater and are more stable than the existing freshwater reagent when used in seawater.

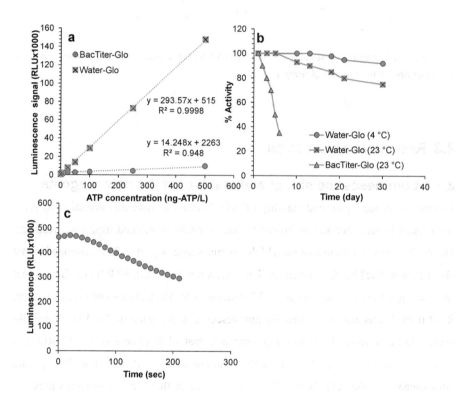

Figure 2.2: (a) The measured luminescence signal of artificial seawater with different ATP standard concentrations ranging from 0 to 500 ng-ATP/L with ATP Water-Glo and BacTiter-Glo reagents. (b) The stability of ATP Water-Glo and BacTiter-Glo reagents over time at different storage temperatures. (c) Stability of luminescence signal over time for a seawater sample measured with ATP Water-Glo reagent.

2.3.2 Effectiveness of the new lysis and detection reagents

The effectiveness of the ATP Water-Glo lysis reagent was tested and compared to the effectiveness of chlorination (8 mg/L) with respect to cell lysis (Fig. 2.3a). The measured microbial ATP concentrations when using the ATP water-Glo lysis reagent (120 ng-ATP/L) and 8 mg/L free chlorine (115 ng-ATP/L) were very similar indicating that the lysis efficiency of Water-Glo lysis reagent is highly effective. A strong lysis reagent may lyse algal cells as well as bacterial cells. However, this has no influence on the measurement of bacterial growth potential as the sample is pasteurised and incubated in the dark.

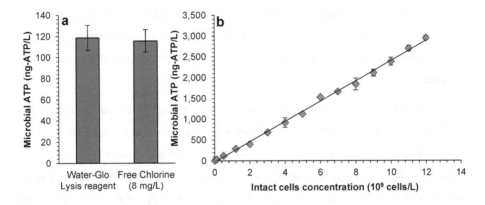

Figure 2.3: (a) Comparing lysis efficiency of ATP Water-Glo and free Chlorine in a raw seawater sample collected from Kamperland, The Netherlands. (b) Measured microbial ATP and intact cell concentration (measured by FCM) in a concentrated seawater sample with 1.2×10^{10} cells/L. Marine microorganisms of 1 L seawater were concentrated in 30 mL.

The results of microbial cell concentrations in raw seawater that are effectively lysed by 100 µL of lysis reagent are presented in Figure 2.3b. A linear relationship was observed (4×10^3 cells/mL = 1.0 ng-ATP/L) between the intact cell concentration measured by flow cytometry and the microbial ATP concentration up to 1.2×10^{10} intact cells/L. This relation suggests that the use of 1: 1 ratio of (seawater sample to Water-Glo lysis reagent) allows measurement of microbial ATP up to 3,000 ng-ATP/L, which is equivalent to an AOC of

2,000 µg–glucose/L, based on a bacterial yield factor determined in our lab for North seawater (Chapter 4).

According to the manufacturer, the volume of Water-Go detection reagent should be equal to or greater than the (combined) volume of the seawater sample and lysis reagent. To demonstrate that sufficient Water-Glo detection reagent volume was used in this study, microbial ATP concentration was measured employing 3 different ratios of seawater sample volume: lysis reagent volume: detection reagent volume (100:100:100 µL, 100:100:200 µL and 100:200:300 µL). In the first ratio, half of the recommended volume (100 µL) of detection reagent (according to the manufacturer) was used whereas, in the second and third ratios, the recommended volume of detection reagent (200 and 300 µL) was tested. To convert the RLU signal into ATP concentration, calibration lines (Fig. 2.4a) for both free and total ATP were prepared for each test, considering the same total volume.

Microbial ATP concentration for the three different volumes of detection reagent (100, 200 and 300 µL) were similar (352 ± 6.2 ng-ATP/L) as shown in Figure 2.4b. From the obtained results for the 100:100:100 µL and 100:100:200 µL volume ratios (Fig. 2.4b, first two columns), it appears that the recommended volume is sufficient to measure all microbial ATP in the seawater samples. Moreover, the results of the 100:100:200 and 100:200:300 µL volume ratios (Fig. 2.4b, last two columns) suggest that a high volume of detection reagent does not impact the final concentration. In conclusion, similar microbial ATP concentrations with different volumes of Water-Glo detection reagent demonstrates that the use of 100 - 300 µL of detection reagent volume was sufficient to detect all of the microbial ATP in the seawater samples.

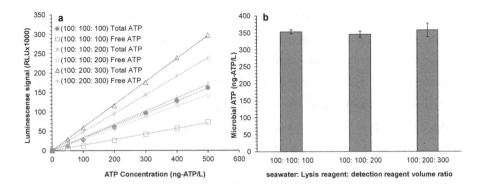

Figure 2.4: (a) Calibration lines of the tested sample sets prepared by diluting ATP standard (100 nM ATP standard) and (b) Measured microbial ATP concentration of seawater with different volume ratio of ATP Water-Glo lysis reagent and ATP Water-Glo detection reagent.

2.3.3 Effect of pH and iron concentration on the luminescence

Since the ATP lysis and detection reagents are added directly to seawater, the chemical composition of the seawater sample may affect the measured ATP concentration. Several studies showed that pH, magnesium concentration, and temperature play a role in the determination of ATP concentration (Hubley et al. 1996, Bergman et al. 2010, Ma et al. 2012).

In this research, the effect of seawater pH and iron concentration present in the sample have been investigated, since both acid and iron based coagulants are commonly applied in desalination plants.

High variations in the luminescence signal were observed at different seawater pH values in which the maximum luminescence signal was at seawater pH 8 – and the signal reduced by 40 % and 60 % at pH 7 and 8.5 (Fig. 2.5a) respectively. The variations may fluctuate depending on the buffering capacity of seawater. In SWRO plants, the pH of seawater is expected to decrease to less than 8 through the pre-treatment, depending on whether acid or coagulant is dosed and the respective concentration of each.

37

Similarly, the luminescence signal decreased (Fig. 2.5b) when the iron concentration in the seawater increased. For example, when the iron concentration increased from 0.1 to 10 mg-Fe(III)/L, the signal decreased by 62 %. However, the iron concentration in the feedwater of the coagulation system of an SWRO plant is not expected to exceed 0.05 mg/L (maximum concentration recommended by the membranes supplier of DOW and Hydranautics). These results show that seawater pH and iron concentration affect the luminescence signal, and in turn ATP measurement.

In full-scale SWRO plants, seawater characteristics change through the pre-treatment train, which will affect the luminescence signal. For example, if the pH in the pre-treatment is decreased as a result of acid dosing, the measured luminescence signal of the seawater sample will be less than the actual signal – which in turn underestimates the measured microbial ATP concentration. Therefore, to eliminate the luminescence signal variation caused by the differences in seawater characteristics, it is important to prepare different calibration lines taking these differences into consideration.

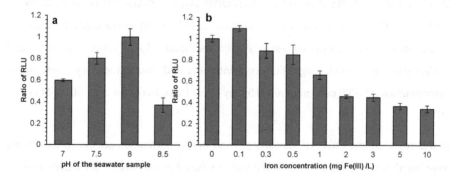

Figure 2.5: (a) The effect of (a) pH of the seawater sample and (b) iron concentration present in seawater on the luminescence signal.

2.3.4 Calibration and limit of detection determination
▪ Calibration line

Seawater characteristics (salinity, pH, etc.) play a significant role in the emitted luminescence signal as discussed earlier. Thus, to calculate ATP accurately, it is important to prepare a calibration curve with similar properties to the real seawater samples.

To investigate the optimum representative calibration line, the slope and intercept of different calibration lines were studied and compared with ATP standard addition to real seawater. These calibration lines were prepared with artificial seawater, pasteurized seawater (70 °C), sterilized seawater (121 °C) and filtered seawater (0.1 μm). It was found that the slopes of all calibration lines were very similar (Table 2.2 and Fig. 2.6) ranging from 557 to 560 RLU.L/ng-ATP which demonstrates that all tested seawater samples (treated with filtration or autoclaving) have similar characteristics to real seawater (without any treatment). The high intercept (y-axis) values for real seawater, pasteurized seawater and filtered seawater calibration lines (65,365; 10,611; and 5,996 RLU, respectively) are due to total/ free ATP concentration present in the sample. This result suggests that both sterilized seawater and artificial seawater may be used to calibrate microbial ATP in seawater since their slopes were similar to real seawater and their background levels are very low (intercept with y- axis) (Fig. 2.6). However, preparing artificial seawater with similar properties to real seawater is very tedious. Therefore, the use of sterilized seawater at 121 °C is recommended.

Since it could be tedious to prepare several calibration lines as seawater characteristics may change along the pre-treatment processes. It is suggested to apply this method for monitoring of a sample over time, such as the determination of bacterial growth potential and AOC concentration and the monitoring of raw seawater. However, it can be applied for any seawater application as long as the calibration line represents the characteristics of the seawater sample (pH, iron concentration, etc.).

Table 2.2: Calibration curves prepared in real seawater, pasteurized seawater, sterilized seawater, filtered (0.1 µm) seawater and artificial seawater.

Calibration line properties		Real seawater - standard addition	Pasteurized seawater - (70 °C)	Sterilized seawater - (121 °C)	Filtered seawater - (0.1 µm)	Artificial seawater
Calibration line	Slope of the calibration (RLU.L/ng)	559.9	559.5	556.7	557.9	558.4
	Regression coefficient (R^2)	0.998	0.992	0.999	0.996	0.999
Intercept point with y- axis	Average (RLU)	65,365	10,611	661	5,996	516
	Standard deviation (RLU)	742	82	20	249	23
	Variation coefficient (%)	1.2	1.5	4	4.2	4.5

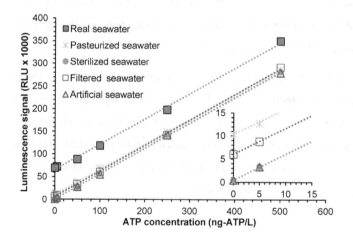

Figure 2.6: Calibration lines prepared in real seawater, pasteurized seawater, sterilized seawater, filtered seawater and artificial seawater with standard addition of ATP ranging from 0 to 500 ng-ATP/L.

- **Limit of detection**

The limit of detection (LOD) was investigated for the recommended volume ratio (100 µL of seawater sample: 100 µL of Water-Glo lysis reagent: 200 µL of Water-Glo detection reagent). The LOD for total ATP and free ATP were 0.2 and 0.2 ng-ATP/L, respectively (Table 2.3) (Taverniers et al. 2004). The combined LOD of microbial ATP was 0.3 ng-ATP/L, which is approximately 1,200 cell/mL (using the correlation shown in Fig 2.3b). The reported LOD of freshwater ATP methods are in the range between 0.05 and 5.1 ng-ATP/L (Velten et al. 2007, Hammes et al. 2010b, Liu et al. 2013b, Vang et al. 2014).

As this method is intended for monitoring bacterial growth potential in SWRO plants, the method should be able to measure the lowest expected concentration in SWRO feed water. However, there is no threshold concentration recommended for ATP in seawater in the literature. Thus, the threshold concentration for AOC in freshwater was used instead. Hijnen et al. (2009a) reported the lowest threshold concentration of AOC to avoid biofouling in freshwater RO system (1 µg-acetate/L), which is approximately 1×10^4 cells/mL (Hammes et al. 2010a). The LOD of the direct ATP method in seawater (0.3 ng-ATP/L, 1,200 cells/mL) is approximately 8 times lower than the reported threshold concentration (1×10^4 cells/mL) suggesting that ATP can be used to monitor bacterial growth potential in SWRO feed water.

Table 2.3: The calculated limit of detection of total ATP and free ATP (n = 10).

	Total ATP	Free ATP
Average blanks (RLU)	584	191
Standard deviation (RLU)	16	11
LOD (ng-ATP/L)	0.2 ± 0.1	0.2 ± 0.1
LOD of the method	0.3	

2.3.5 Application of the microbial ATP method

- **Microbial ATP monitoring of raw seawater**

The microbial ATP concentration of seawater (The Netherlands) was regularly monitored (weekly to bi-weekly) over 2016. The concentration ranged from 25 ng-ATP/L to 1,037 ng-ATP/L with the lowest concentration observed during the winter months as shown in Fig. 2.7. It is not unlikely that micro-algae were also lysed by the Water-Glo lysis reagent. In the winter, microbial ATP ranged between 20 ng-ATP/L and 140 ng-ATP/L with an average of 52 ng-ATP/L. During the spring period, the microbial ATP concentration started to increase above 100 ng-ATP/L and reached 1,000 ng-ATP/L. This increment could be due to the ATP extraction from the algal cells or due to the presence of sufficient nutrients released from algal cells during the bloom period which led to high bacterial growth. An algal bloom was noticed in April, when the algal cell counts increased from 10 to 1,000 cells/mL. Another possible reason would be due to the higher activity of microorganisms at higher temperatures. After the spring season, the microbial ATP concentration declined to a range below 100 ng-ATP/L. The variation of marine microbial ATP over time may indicate fluctuations in the amount of nutrients in seawater.

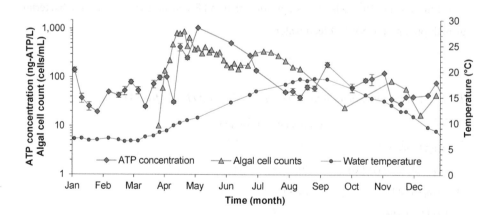

Figure 2.7: Microbial ATP and algal cell concentrations in raw seawater collected from Kamperland (North Sea) between January and December 2016. All data points are plotted as average ± standard deviation (n = 3 each).

In the measured samples from the North Sea, the percentage of microbial ATP ranged between 55 and 88 % of the total ATP with an average of 74 ± 9 % (Fig. 2.8). Free ATP is not marginal and accounted

(> 12 % of the total ATP) which reveals that the use of total ATP for indicating the microbial activity in the seawater instead of microbial ATP may be misleading. Moreover, the abundance of microbial ATP and free ATP could vary depending on the seawater sample used and the bacterial growth phase.

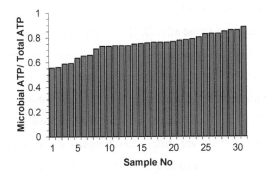

Figure 2.8: The fraction of microbial ATP over total ATP of seawater from the North Sea (Kamperland, The Netherlands). All data bars are plotted as average (n = 3).

- ## Monitoring of microbial ATP and bacterial growth in an SWRO desalination plant

The new reagents were used to measure the microbial ATP concentration in a full-scale membrane-based desalination plant. Results showed the maximum microbial ATP concentration (90 ng-ATP/L) in the raw seawater of Australian SWRO desalination plant (Fig. 2.9), which is relatively low compared to the measured microbial ATP concentration in the North Sea during the spring (300-1,000 ng-ATP/L). The ATP concentration gradually decreased through the pre-treatment processes from 90 to 55, 38, and 19 ng-ATP/L after flocculation, dual media filter and after cartridge filtration, respectively. Different calibration lines were established for each sample due to the changes of the seawater matrix across the pre-treatment (pH, iron, magnesium, etc.), as presented earlier.

Microbial ATP measurement was also applied to monitor the bacterial growth potential across the pre-treatment processes of an SWRO plant (Fig. 2.10a). After bacterial inactivation of the seawater samples, the samples were inoculated with an average microbial ATP concentration of 7.8 ± 1.7 ng-ATP/L. Bacteria started to grow immediately in seawater and reached a maximum growth within 2 days. Afterwards, microbial ATP gradually decreased, either due to the partial decay of bacteria or because bacterial activity decreased due to the depletion of nutrients. As expected, the maximum bacterial growth was observed (305 ng-ATP/L) in raw seawater (Fig. 2.10b), indicating the highest potential of bacterial growth. Slightly lower potential of bacterial growth (262 ng-ATP/L) was noticed after coagulation and flocculation, while a significant reduction (> 55 %) of the bacterial growth potential was found after DMF – therefore indicating a biologically-active media filter. This high removal in the DMF coincided with the reported removal by Weinrich et al. (2011) in which the removal in the sand filtration ranged between 25 and 70 %. The maximum bacterial growth decreased modestly through the cartridge filter to 86 ng-ATP/L. This result shows that the seawater after pre-treatment still supports further bacterial growth as there are differences between the present microbial ATP (19 ng-ATP/L) and the maximum microbial ATP (86 ng-ATP/L) that can be found in the tested seawater. It should be noted that the protocol determination of bacterial growth potential based on microbial ATP measurements using indigenous microbial culture will be discussed in depth in a following article.

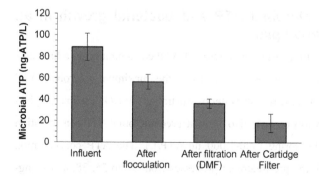

Figure 2.9: Microbial ATP through the RO pre-treatment processes of an SWRO desalination plant in Australia (n=3).

The monitored microbial ATP and bacterial growth potential based on microbial ATP illustrate the applicability of the new developed reagents for measuring microbial ATP and that this method can be applied to measure bacterial growth potential in seawater.

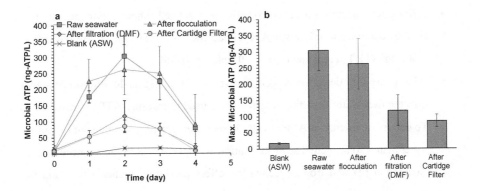

Figure 2.10: (a) bacterial growth over time and (b) maximum growth of different seawater samples collected through the pre-treatment processes of an SWRO desalination plant in Australia. Bacteria in the samples were inactivated and then inoculated with 7.8 ± 1.7 ng-ATP/L of microbial ATP.

2.4 CONCLUSIONS

- The applicability of new reagents (Water-Glo lysis and detection reagent) to measure microbial ATP directly in seawater has been demonstrated.

- Water-Glo lysis reagent shows strong lysis efficiency (similar to 8 mg/L free chlorine) in seawater.

- A linear relationship was observed between intact cell concentration measured by flow cytometry and microbial ATP concentration in seawater in the range 0 to 3,000 ng-ATP/L (equivalent to 1.2×10^7 intact cells/mL).

- ATP Water-Glo detection reagent showed 20 times higher luminescence signal than the freshwater detection reagent, when used to measure ATP in seawater.

- To determine microbial ATP directly in seawater, a calibration line with a similar water matrix to the actual seawater sample is required. Calibration is necessary as changes in pH and iron concentration affect the luminescence signal and the measured ATP concentration.

- The limit of detection of the direct method to determine microbial ATP in seawater is 0.3 ng-ATP/L (equivalent to 1,200 cell/mL).

- Microbial ATP concentration in North Sea has been monitored and high seasonal variations were observed ranging from 20 ng-ATP/L to 1,000 ng-ATP/L.

- Microbial ATP has been applied to measure bacterial growth potential using an indigenous bacterial consortium in an SWRO desalination plant in Australia. A significant reduction (55 %) in bacterial growth potential was noticed through dual media filtration with 4.5 mg-Fe^{3+}/L coagulant added prior to dual media filtration.

- Ongoing research will focus on the applicability of microbial ATP for monitoring bacterial growth potential in SWRO plants around the world.

2.5 ACKNOWLEDGMENTS

Special thanks goes to David Grasso from Australia for access to the SWRO plant. This study was made possible by funding from the Promega Corporation (Madison, Wisconsin, USA).

3

ELIMINATING SEAWATER MATRIX EFFECTS IN ATP MEASUREMENT USING A FILTRATION PROCESS

A direct method for measuring adenosine-triphosphate (ATP) in seawater was developed recently, in which commercial reagents are added directly to seawater. However, calibration is required if seawater quality changes (such as changes in salinity, pH, Mg^{2+}, Fe^{3+}) as the seawater matrix interferes with ATP measurement. In this chapter, a 0.1 μm filtration process is introduced to eliminate such interferences. In addition, a filter rinsing step with sterilized artificial seawater is proposed to eliminate interference of free ATP.

The ATP-filtration method is fast (< 5 min), reproducible (VC = 7 %), six times more sensitive than the direct ATP-method and correlates ($R^2 = 0.72$, n = 100) with intact cell concentration. Microbial ATP concentration measured using the ATP-filtration method and the ATP-direct method were comparable. Microbial ATP measured along the treatment train of a full-scale seawater reverse osmosis (SWRO) plant decreased from 530 in the raw seawater to 10 ng-ATP/L after pre-treatment and to 0.5 ng-ATP/L in the SWRO permeate. The method was also applied to monitor bacterial growth potential (BGP) across the pre-treatment train of a (pilot) seawater desalination plant, where the removal of BGP through the media filtration and ultrafiltration was 44 % and 7 %, respectively.

This chapter has been published as **Almotasembellah Abushaban**, Sergio. G. Salinas-Rodriguez, M. Nasir Mangal, Subhanjan Mondal, Said A. Gouel, Aleksandra Knezev, Johannes S. Vrouwenvelder, Jan C. Schippers, Maria D. Kennedy (2019) ATP measurement in seawater reverse osmosis systems: eliminating seawater matrix effects using a filtration-based method. *Desalination* (2019) 453 p 1–9.

3.1 INTRODUCTION

In reverse osmosis (RO) desalination, microbial quantification has been implemented: (*i*) to quantify biomass accumulation on RO membranes for biofouling diagnostics (Vrouwenvelder et al. 1998, Vrouwenvelder et al. 2008), (*ii*) to measure biomass in the feed water and across pre-treatment trains in RO plants (Vrouwenvelder and Van der Kooij 2001, Veza et al. 2008), (*iii*) as a biomass parameter for the determination of nutrients (carbon) (Van der Kooij et al. 1982, Stanfield and Jago 1987, Farhat et al. 2018), and (*iv*) to measure bacterial growth potential (Stanfield and Jago 1987, Withers and Drikas 1998, Van der Kooij and Van der Wielen 2013).

The common methodologies to quantify microbes are heterotrophic plate counts (HPC's) and total direct counts (TDC's). HPC has been used to monitor microbial populations in seawater (Jannasch and Jones 1959), drinking water treatment (Bartram et al. 2003) and distribution systems (WHO 2006) but it is a laborious and time consuming method. HPC is also limited to the enumeration of cultivable bacteria, which often comprise less than 1 % of the active bacterial population in natural water (Staley and Konopka 1985, Liu et al. 2013b). TDC enumerates the total numbers of cells but does not distinguish between active and inactive cells and is limited to samples that have high cell concentrations ($>10^7$ cell/mL) (Postgate 1969). To avoid these limitations, flow cytometry (FCM), and adenosine triphosphate (ATP) have been proposed as alternative methods. FCM offers rapid enumeration of the total number of bacterial cells in water using bacterial DNA staining. ATP is the energy source in living organisms and it is used a measure for the amount of active biomass (Karl 1980, Knowles 1980, Webster et al. 1985). ATP and FCM have attracted increasing interest because such methods are considered to be more accurate, rapid, and quantitative, can detect both cultivable and non-cultivable microorganisms, can be automated and are easy to perform (Vang et al. 2014, Van Nevel et al. 2017).

In freshwater aquatic environments, ATP has been used to monitor water quality across treatment trains in drinking water plants (Berney et al. 2008, Hammes et al. 2008, Siebel et al. 2008), measuring active biomass on granular activated carbon, sand and anthracite grains (Magic-Knezev and Van der Kooij 2004), biostability of drinking water and biofilm formation in distribution systems (Boe-Hansen et al. 2002, Liu et al. 2013a, Prest et al. 2016) and determining ATP in ground water (Ludvigsen et al. 1999, Eydal and Pedersen 2007). ATP in aquatic systems can be found within live/active microorganisms (microbial ATP) or ATP in the water, which has been released from dead or stressed living microorganisms (free ATP). Various studies have been conducted to evaluate microbial activity in freshwater either using total ATP (microbial ATP and free ATP) or only microbial ATP (Magic-Knezev and Van der Kooij 2004, Vrouwenvelder et al. 2008, Hammes et al. 2010b). However, the application of ATP in seawater is limited due to the interference of salts in the luciferase-luciferin reaction. The high concentration of salt in seawater has been demonstrated by Van der Kooij and Veenendaal in Amy et al. (2011) to cause substantial reduction of the emitted light during the enzymatic ATP luciferase-luciferin reaction at a high salt concentration (>10 g/L).

The standard method for microbial ATP determination in all types of water used by the American Society for Testing and Materials (ASTM) includes a filtration process (0.45 μm pore size) in which the filter is transferred and placed in tris-buffer after filtration (ASTM Standard D 4012 1981(Reapproved 2002)). The filter and the tris-buffer are heated at 100 °C to extract the microbial ATP from microorganisms. This method is complicated, and contamination may be introduced while handling the measurement. Van Slooten et al. (2015) followed the same concept outlined by ASTM to quantify microbial ATP concentrations of living organisms (with a size between 10 and 50 μm) in ballast water using a 10 μm pore sized nylon filter. Marine microorganisms captured on the filter were flushed out with Milli-Q water to eliminate interference from salt, and then the filter, including the retained microorganisms, was placed into a cuvette with 2 mL of Milli-Q water before analysing microbial ATP using freshwater reagents. This method is also complicated, and the use of Milli-Q water to flush marine microorganisms can rapidly burst bacterial cells due to osmotic shock.

Recently, direct quantification of microbial, free, and total ATP (total ATP is the sum of microbial ATP and free ATP) determination in seawater was proposed by Abushaban et al. (2018) using new commercial reagents (Water-Glo kit, Promega, USA). The new reagents overcome the luciferin-luciferase problem and provide a high luminescence signal even in the presence of salt. The reagents are added directly to seawater for microbial ATP extraction and detection. The method is simple, direct, allows ATP determination at a low concentration level (limit of detection (LOD) = 0.3 ng-ATP/L) and is promising for monitoring microbial growth potential in SWRO systems. Nevertheless, a calibration line is needed each time the seawater quality changes (i.e., pH, total dissolved salt (TDS), Mg^{2+}, Fe^{3+}, turbidity) affect the ATP measurement. Preparing several calibration lines may be tedious in some applications, such as along the pre-treatment train of an SWRO plant since pH, Fe^{3+}, turbidity, etc. vary according to the applied settings of each treatment step.

In this chapter, the interference of salt in microbial ATP determination was eliminated using a filtration step. Filtration allows the capture of marine microorganisms on a filter surface. Thus, eliminating the seawater matrix effect. To remove free ATP present in the filter holder after filtration, the captured microorganisms on the filter surface are rinsed with sterilized artificial seawater. Moreover, the use of filtration can improve the detection limit of the method by increasing the sample volume, enabling measurement of samples that have low ATP concentrations (such as after microfiltration and ultrafiltration). The following aspects have been addressed in this chapter:

- Pore size of the filter and flushing of free ATP from the filter holder.
- Method properties, including the limit of detection, reproducibility, and the correlation with intact cell counts measured by FCM in North Sea water.
- Comparing ATP measurement with the ATP-direct method and the current ATP-filtration method.
- Monitoring bacterial growth potential in seawater.
- Measuring microbial ATP concentrations across the pre-treatment train of a full-scale SWRO desalination plant.

3.2 MATERIALS AND METHODS

3.2.1 Sample collection and storage

Natural coastal seawater samples from North Sea water (Kamperland, The Netherlands) were collected from September 2017 to May 2018 in sterile 500 mL glass sampling bottles with tight-fitting screw caps. The samples were transported (110 km) for analysis to IHE Delft (Delft, The Netherlands) in a cooler (5 °C). The characteristics of the collected seawater samples are: pH = 7.9 ± 0.1, TDS = 32.5 ± 0.8 g/L, electrical conductivity = 52.6 ± 1.2 mS/cm, TOC = 1.28 ± 0.85 mg/L, UV_{254} = 0.045 ± 0.009 1/cm, total bacterial count measured with FCM = 0.9 ± 0.28 $\times 10^6$ cells/mL.

3.2.2 Preparation of ASW

Artificial seawater containing all ions (ASWall ions) in seawater was prepared using Milli-Q water (Milli-Q® water Optimized purification, 18.2 MΩ.cm at 25°C, Millipore, USA) and analytical or reagent-grade inorganic salts (Merck, USA) with similar ion concentrations as the average global seawater (Villacorte 2014) (23.668 g/L NaCl, 10.873 g/L $MgCl_2.6H_2O$, 3.993 g/L Na_2SO_4, 1.54 g/L $CaCl_2.2H_2O$, 0.739 g/L KCl, 0.213 g/L $NaHCO_3$, and 0.002 g/L Na_2CO_3). Similarly, ASW $_{NaOH + NaHCO3}$ was prepared using 33.2 g/L NaCl and 0.213 g/L $NaHCO_3$. All salts except sodium bicarbonate were mixed (150 rpm) with Milli-Q water for 24 hours and autoclaved at 100 °C for 20 minutes. Sodium bicarbonate was added after autoclaving because its melting point is 50 °C.

3.2.3 Microbial ATP measurement

Figure 3.1 presents the protocol employed in the ATP-filtration method. Seawater samples (5 mL) were filtered through disposable sterile 0.1 µm PVDF membrane filters (Millex GP, Merck Millipore, USA). Two millilitres of sterilized ASW$_{all ions}$ were filtered through the same filter to flush out the remaining volume of seawater in the filter holder because to ensure complete removal of free ATP. Afterwards, 5 mL of the Water-Glo lysis reagent (Promega Corp., USA) was added to extract microbial ATP from the captured microorganisms, and then the filtrate was collected in 15 mL sterile centrifuge tubes. The filtrate and the ATP Water-Glo detection reagent were separately and simultaneously heated to 38 °C in a dry heating block (AccuBlock™ Digital, Labnet,

USA). A 100 μL aliquot of the filtrate was added to 100 μL of the ATP Water-Glo detection reagent, and then the luminescence generated was measured with a GloMax®-20/20 instrument (Promega Corp.). The filtration rate and pressure were tested for all filtration steps and found that the filtration rate (range: 50 to 400 L/m.hr) had no effect on the measured microbial ATP concentration.

The measured emitted light in relative light units (RLU) was converted to a microbial ATP concentration based on a calibration line. To prepare a calibration line, the procedure described above was followed but without seawater sample filtration. The filtrate of the Water-Glo lysis reagent and standard ATP (1,000 ng-ATP/L, Promega Corp., USA) were used to prepare standard ATP solutions ranging between 0 and 500 ng-ATP/L.

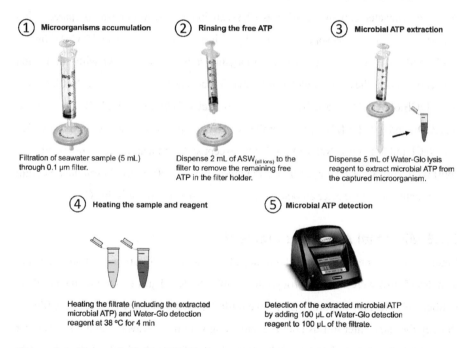

① Microorganisms accumulation

Filtration of seawater sample (5 mL) through 0.1 μm filter.

② Rinsing the free ATP

Dispense 2 mL of ASW(all ions) to the filter to remove the remaining free ATP in the filter holder.

③ Microbial ATP extraction

Dispense 5 mL of Water-Glo lysis reagent to extract microbial ATP from the captured microorganism.

④ Heating the sample and reagent

Heating the filtrate (including the extracted microbial ATP) and Water-Glo detection reagent at 38 °C for 4 min

⑤ Microbial ATP detection

Detection of the extracted microbial ATP by adding 100 μL of Water-Glo detection reagent to 100 μL of the filtrate.

Figure 3.1: Protocol of the ATP-filtration method to measure microbial ATP in (sea) water.

To compare the measured microbial ATP using the ATP-filtration method with the measured microbial ATP using the ATP-direct method, 125 seawater samples from North Sea water and Tasman seawater were tested using both protocols. The procedure of the ATP-direct method in seawater, described in Abushaban et al. (2018), was followed. Figure 3.1 depicts the protocol that has been followed to monitor microbial ATP concentration using the ATP-filtration method.

3.2.4 Tested variables in the ATP-filtration method

Several variables that might affect the performance of the ATP-filtration method were tested, including the filter pore size, rinsing free ATP from the filter holder, and the effect of the seawater sample volume.

There is evidence that a significant amount of marine bacteria could pass through membrane filters with a pore size of 0.45 μm and even through 0.2 μm (Macdonell and Hood 1982, Denner et al. 2002). We compared 0.1 μm PVDF sterilized filters, 0.22 μm PES sterilized filters and 0.45 μm PVDF sterilized filters (Millex GP, Merck Millipore, USA) based on the retained microbial ATP concentration (measured using the ATP-filtration method) on the filter surface. The properties of the used filters are shown in Table 3.1.

To test the effect of rinsing the filter on removal of free ATP from the holder, different options were tested, including: (1) no rinsing, (2) rinsing with 2 mL of demineralized water, (3) rinsing with 2 mL of ASW comprising sodium, chloride, and bicarbonate ions (ASW$_{NaCl+NaHCO3}$), and (4) rinsing with 2 mL of ASW containing all the major ions in seawater (ASW$_{all\ ions}$).

The effect of the seawater sample volume was also studied by measuring the luminescence signal and the calculated microbial ATP of different sample volumes ranging between 1 and 30 mL for seawater samples.

Table 3.1: Properties of the tested membrane filters (Merck Millipore 2012).

Pore size	Millex –VV syringe filter, 0.1 μm	Millex –GP syringe filter, 0.22 μm	Millex –HV syringe filter, 0.45 μm
Sterility	Sterile	Sterile	Sterile
Material	Polyvinylidene Fluoride (PVDF)	Polyethersulfone (PES)	Polyvinylidene Fluoride (PVDF)
Wettability	Hydrophilic	Hydrophilic	Hydrophilic
Maximum inlet pressure	10 bar	10 bar	10 bar
Bubble point	73 psi (5.1 bar)	57 psi (3.9 bar)	25 psi (1.7 bar)
Filter diameter	33 mm	33 mm	33 mm
Filtration area	4.5 cm^2	4.5 cm^2	4.5 cm^2

3.2.5 Total intact cell counting using flow cytometry (ICC-FCM)

Intact cells were counted by double DNA staining and flow cytometric analysis as described elsewhere (Van der Merwe et al. 2014). In brief, 10 μL of the SYBR green solution was mixed with 10 μL of Propidium iodide (1 mg/mL) and 980 μL of 0.22 μm filtered DMSO solution. SYBR green is a cell-permeable DNA binding dye and can bind to DNA of either intact or damaged cells, while Propidium iodide is a membrane-impermeable DNA binding dye and binds to DNA in cells that have only lost their membrane integrity. Simultaneous staining with SYBR green and Propidium iodide allows a distinction to be made between intact and damaged bacteria. The water sample (500 μL) was first heated to 36 °C for 5 minutes and then stained by adding 5 μL of the SG/PI solution. The stained sample was then incubated at 36 °C for 10 minutes. Five hundred microlitres of the stained sample were injected at medium speed into the flow cytometer (BD Accuri C6). The result was visualized in a special gate designed for seawater samples and counted using a two-dimensional FL1-A (emission filter 533/30) vs. a FL3-A (emission filter 670 LP) log-scale density plot. The range of bacterial counts can be as low as 100 cell/mL and as high as 10^7 cells/mL.

3.2.6 Comparing ATP-filtration and ATP-direct methods

The ATP-filtration method was compared to the ATP-direct method because it is the only method applied in seawater using the same reagents. The comparison was made based on the microbial ATP concentration in 2 different applications:

(1) In a full-scale desalination plant, seawater samples are collected along the treatment train of an SWRO desalination plant. The treatment line of the plant included a drum screen, coagulation, dual media filtration, cartridge filter, 2 pass RO membrane, and remineralisation.

(2) Several locations: 125 raw seawater samples were collected from the Tasman Sea (Australia) and the North Sea (The Netherlands).

3.2.7 Monitoring of bacterial growth potential in a seawater pilot plant

Seawater samples were collected from raw seawater after a media filter (AFM® Active Filter Media, Grade 2, RD397) and after ultrafiltration from a pilot-scale plant (Kamperland, The Netherlands). The samples were pasteurized at 70 °C for 1 hour and inoculated with 10,000 intact cells per mL (measured by flow cytometry) using an indigenous microbial inoculum from the same seawater source. The samples were incubated at 30 °C, and microbial ATP concentrations were monitored daily using the ATP-filtration method to assess the bacterial growth potential.

3.2.8 Statistical analysis

The average of triplicate measurements was reported, and the standard deviation was shown as the positive and negative error bars. The linear regression and Spearman correlation were calculated to assess the correlation between microbial ATP and intact cell concentration and the correlation between the ATP-filtration method and ATP-direct method. Moreover, an analysis of variation (ANOVA) was used to determine the significance of the correlation, including a P-value test and t-test of two samples assuming equal variances.

3.3 RESULTS AND DISCUSSION

3.3.1 Selection of filter pore size

Filter pore size may play a significant role in the determination of microbial ATP of each method. As mentioned earlier, ASTM method uses a 0.45 µm filter to quantify microbial ATP in water. However, Bowman et al. (1967) reported that smaller organisms were not retained in a 0.45 µm filter (Meltzer and Jornitz 2006), and Macdonell and Hood (1982) observed that smaller marine bacteria (*Bdellovibrio*) in the Gulf of Mexico can even pass through 0.2 µm filters. To verify whether small size bacteria significantly influence the measured microbial ATP, a comparison of three filter pore sizes (0.1 µm, 0.22 µm and 0.45 µm) was performed for North Sea water. Higher microbial ATP concentrations were found using a smaller filter pore size for all seawater samples measured over 8 months, which ranged between 265 and 1,335 ng-ATP/L (Fig. 3.2). The microbial ATP measured using a 0.22 µm filter was 12 to 47 % lower than using a 0.1 µm filter. Furthermore, the microbial ATP measured using a 0.45 µm filter was 16 to 50 % lower than the microbial ATP measured using a 0.1 µm filter. The variation in these percentage could be due to seasonal variations of microbial ATP or the prevalence of certain seasonal bacterial species smaller than 0.22 µm. At the beginning of spring (March and April), a high microbial ATP concentration was captured on 0.1 µm filter compared to the captured microbial ATP concentration on the 0.22 and 0.45 µm filter, which could be due to the algal bloom and the presence of microalgae.

The retention of microorganisms on a 0.1 µm filter was tested by measuring intact cell concentration in the raw seawater and the filtrate. In average, more than 99.9 % of microorganism were retained on the 0.1 µm filter (See Table S3.1). Accordingly, the 0.1 µm filter was selected for the filtration process. This conclusion is in agreement with the findings of Wang et al. (2007) and Hammes et al. (2008) in which freshwater bacteria were detected in the filtrate of the 0.22 µm filter and thus used the 0.1 µm filter to distinguish between microbial ATP and free ATP in freshwater. This conclusion was drawn without testing the effect of membrane materials, which needs further study.

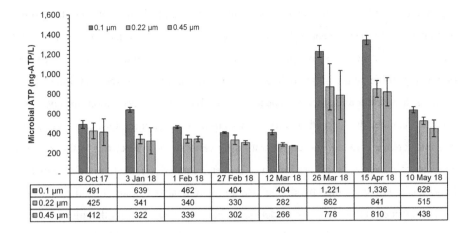

	8 Oct 17	3 Jan 18	1 Feb 18	27 Feb 18	12 Mar 18	26 Mar 18	15 Apr 18	10 May 18
▣0.1 μm	491	639	462	404	404	1,221	1,336	628
▣0.22 μm	425	341	340	330	282	862	841	515
▣0.45 μm	412	322	339	302	266	778	810	438

Figure 3.2: Microbial ATP concentration measured using the ATP-filtration method with 0.1 μm, 0.22 μm and 0.45 μm filter. Seawater samples were collected from the North Sea.

3.3.2 Removal of free ATP

After the seawater sample filtration, a small volume (0.45 mL, < 10 % of the sample) of the seawater remains in the filter holder. The remaining volume also includes free ATP (< 5 % of the total concentration, assuming free ATP is 50 % of total ATP), which interferes with the measured microbial ATP concentration. The interference of free ATP is variable depending on the free ATP concentration in the sample. This retained volume can be removed either by air flushing or rinsing (e.g., with demineralized water). The use of air flushing is impractical with a 0.1 μm filter at a lab scale because the bubble point of a 0.1 μm filter is approximately 5 bar (Merck Millipore 2012). This might be ameliorated by using a pump operating at a constant flow to overcome the bubble point.

To assess the effectiveness of removing free ATP using rinsing, the microbial ATP of the seawater sample was measured and compared using four different rinsing conditions: (1) No rinsing, (2) rinsing with demineralized water, (3) rinsing with ASW comprising sodium, chloride and bicarbonate ions (ASW$_{NaCl+ NaHCO3}$), and (4) rinsing with ASW containing all major ions in seawater (ASW$_{all ions}$).

The microbial ATP measurement without rinsing showed a similar concentration to the microbial ATP measurement with $ASW_{all\ ions}$ (Fig. 3.3) due to the fact that the free ATP concentration in the tested sample was insignificant. Although eliminating the rinsing option appears to be the most obvious solution, the variability of free ATP interference makes it less preferable since it depends on the free ATP concentration present in the seawater sample which could be very significant in some samples. One can suggest that this interference is taken into consideration by calibration. Indeed, this is theoretically applicable; however, a calibration line is then required for each sample, which is impractical. On the other hand, the use of demineralized water to flush seawater (including free ATP in the filter holder) showed a negative effect on the microbial ATP measurement (Fig. 3.3) because the measured concentration was 66 % lower than the microbial ATP concentration measured with other rinsing options. Rinsing marine microorganisms with demineralized water can rapidly burst the captured microorganisms on the filter surface due to osmotic shock, which results in a significant loss of microbial ATP. The survival of marine microorganisms rinsed with different salt concentrations ranging between 0 and 60 g/L was tested, which showed that marine microorganisms (North Seawater) can survive when rinsed with an artificial solution with a salt concentration ranging between 10 and 35 g/L (Figure S3.1).

The microbial ATP concentration using $ASW_{(NaCl\ +\ NaHCO3)}$ was 10 % less than the microbial ATP concentration measured using $ASW_{all\ ions}$ (Fig. 3.3). This difference could be due to the absence of some ions in the $ASW_{(NaCl\ +\ NaHCO3)}$. It was found that calcium and magnesium are essential to the survival of marine microorganisms with regards to cell viability, as shown in the supplementary data Figures S3.2 and S3.3. It should be noted that the use of ASW comprising all ions overcomes the interference of free ATP because $ASW_{all\ ions}$ is sterile and has constant properties. Consequently, it was determined to use $ASW_{(all\ ions)}$ to flush the seawater, which contains free ATP, from the filter holder.

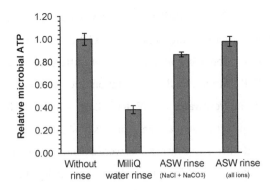

Figure 3.3: Tested rinsing solutions to remove/flush the remaining seawater from the filter holder (including free ATP). Two mL of each solution was applied to remove the free ATP.

3.3.3 Limit of detection and sample volume

The LOD of the ATP-filtration method was calculated based on the average of 10 blanks plus 3 times the standard deviation of the 10 blanks (95 % confidence level). ASW$_{all\ ions}$ was used as a blank sample for LOD determination. The measured LOD of the ATP-filtration method was 0.06 ng-ATP/L based on the proposed volume ratio (1:1, seawater sample: Water-Glo lysis reagent). Abushaban et al. (2018) showed that the use of a 1:1 ratio of the seawater sample to lysis reagent allows measurement of microbial ATP up to 3,000 ng-ATP/L, which was equivalent to about 1.2×10^{10} cells/L (measured by FCM). Additionally, to measure samples below the LOD, a higher volume of seawater can be filtered. For example, to measure a sample with 0.006 ng-ATP/L, the filtered seawater volume needs to be more than 10 times (> 50 mL) that of the lysis reagent volume. By controlling the sample volume, the ATP-filtration method can be more sensitive than the reported LOD of the ATP-direct method (LOD = 0.3 ng-ATP/L) (Abushaban et al. 2018).

The higher sensitivity of the ATP-filtration method is due not only to the sample volume but also to the measured high signal using the Water-Glo reagent because the interference of seawater composition and salts with the reagent was eliminated. The difference between the luminescence signal of the ATP-direct method and the luminescence signal of the ATP-filtration method can be seen in Figure 3.4, in which the calibration lines of both methods were compared. The slope of the ATP-filtration method calibration line (1,293 RLU.L/ng-ATP) was 2.3 times greater than the slope of the ATP-direct method calibration line (563 RLU.L/ng-ATP). Moreover, an insignificant difference was observed between the slope of the ATP-filtration method calibration line in seawater (1,293 RLU.L/ng-ATP) and its slope in freshwater (1,347 RLU.L/ng-ATP), which confirms the deleterious effect of salt on luciferase reaction. This result showed that a higher luminescence signal was obtained when the seawater ions were excluded from the ATP reaction. Theoretically, the high luminescence signal increases the sensitivity of the measurements.

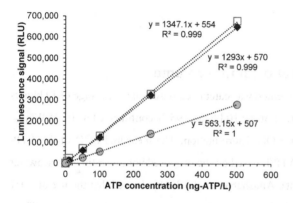

Figure 3.4: Comparison between the calibration curves of the ATP-direct method and the ATP-filtration method in freshwater and seawater. Symbols: (●) ATP-direct method calibration line in seawater, (♦) ATP-filtration method calibration line in seawater, and (□) ATP-filtration method calibration line in freshwater.

The effect of the sample volume was studied by measuring the luminescence signal and the calculated microbial ATP of two seawater samples at different volumes between 1 and 30 mL (Table 3.2 and Figure S3.4). The measured microbial ATP of Sample A ranged between 58.8 and 61.3 ng-ATP/L with 1.5 % coefficient of variation, while the microbial ATP of Sample B ranged between 6.04 and 6.98 ng-ATP/L with 5 % coefficients of variation. This result indicates that the sample volume has no effect on the measured microbial ATP and suggests the use of a higher sample volume for samples with low microbial ATP.

Table 3.2: Measured luminescence signal and calculated microbial ATP of 2 seawater samples at different volumes between 1 and 30 mL.

Sample	Sample A		Sample B	
volume (mL)	Luminescence signal (RLU)	Microbial ATP (ng/L)	Luminescence signal (RLU)	Microbial ATP (ng/L)
1	13,543	58.8	2,424	7.0
2	26,872	59.9	3,845	6.3
3	39,971	59.9	6,034	7.0
5	65,849	59.7	9,082	6.5
10	132,821	60.5	18,683	6.8
15	201,605	61.3	27,362	6.7
20	269,623	61.6	32,712	6.0
30	396,881	60.5	52,280	6.5
Average (ng-ATP/L)		60.3		6.6
Standard deviation (ng-ATP/L)		0.9		0.3
Variation coefficient (%)		1.5		4.5

3.3.4 Correlation with the intact cell concentration

The microbial ATP of 100 seawater samples (North Sea) was measured using the ATP-filtration method, in which the microbial ATP concentrations ranged from 0.5 to 670 ng-ATP/L. For the same set of samples, their intact cell concentration was also measured by flow cytometry (ICC-FCM) and ranged from 1.1×10^3 to 7.3×10^6 cells/mL. The correlation between the two parameters in the seawater is presented in Figure 3.5 (R^2 = 0.72, Rho = 0.88, P-value $\ll$ 0.001, n = 100). A strong correlation was not expected because flow cytometry counts the total number of intact cells regardless of their activity, and microbial ATP measures the activity of cells regardless of their number. A correlation (R^2 = 0.69, n = 200) between the total cell count measured by flow cytometry and total ATP was reported in drinking water (Siebel et al. 2008, Liu et al. 2013b). Furthermore,

Hammes et al. (2010b) reported a roughly similar correlation ($R^2 = 0.8$, P-value $\ll 0.001$, $n = 102$) between microbial ATP and ICC-FCM in freshwater. Van der Wielen and Van der Kooij (2010) also observed correlations ($R^2 = 0.55$ (spring), $R^2 = 0.82$ (winter), P <0.01, $n = 48$) between total ATP and the total count of cells measured with epifluorescence microscopy in drinking water.

Based on the correlation found in this study, the average microbial ATP concentration per cell was estimated at 8.6×10^{-7} ng-ATP/cell. This is in agreement with the findings of Hamilton and Holm-Hansen (1967), who used the ASTM method to measure the ATP content in seven selected cultures of marine bacteria. They reported that the average ATP content ranges from 5 to 65 $\times10^{-7}$ ng-ATP/cell.

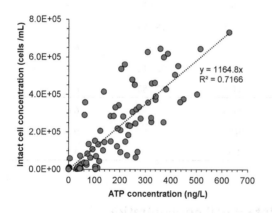

Figure 3.5: Correlation between microbial ATP and intact cell concentration (flow cytometry in seawater ($R^2 = 0.72$, Rho = 0.88, P-value $\ll 0.001$, n = 100). An average microbial ATP per intact cell in seawater was derived from these data (8.6×10^{7} ng-ATP/cell).

3.3.5 Comparing ATP-filtration method with ATP-direct method

Although the same reagents (lysis and detection reagent) were used to measure seawater in both the ATP-filtration and ATP-direct method, the protocol of each method is different. The main difference is that in the ATP-direct method, the reagents are added directly to the seawater; therefore, the matrix effects of the sample need to be taken into consideration by preparing a calibration line. In the ATP-filtration method, the microorganisms are collected on a membrane filter, and the microbial ATP is extracted on the filter itself. Table 3.3 summarizes the similarities and differences between the ATP-filtration method and the ATP-direct method.

Table 3.3: Overview comparison between ATP-direct method and ATP-filtration method.

	ATP-direct method	ATP-filtration method
Reagent for microbial ATP extraction	Water-Glo lysis reagent	Water-Glo lysis reagent
Reagent for light generation	Water-Glo detection reagent	Water-Glo detection reagent
Measurement	Microbial ATP = Total ATP - Free ATP	Microbial ATP only
Microbial ATP extraction	Direct - in the seawater sample	On a filter surface
Complexity	Simple (2 steps)	Complex (4 steps)
Cost	Low	Moderate
Limit of detection	0.3 ng-ATP/L	< 0.06 ng-ATP/L
Matrix effect	Yes (pH, Mg^{+2}, Fe^{+2})	No

- **Pre-treated seawater in an SWRO plant**

The microbial ATP concentration was measured through the pre-treatment train of the SWRO desalination plant using the ATP-filtration and ATP-direct methods (Fig. 3.6). In general, microbial ATP concentrations of both methods were comparable and ranged from 19 to 89 ng-ATP/L. Similar microbial ATP concentrations were measured (80 and 89 ng-ATP/L) in raw seawater using the ATP-filtration method and the ATP-direct method, respectively. A slight reduction (20 %) in microbial ATP was noted after flocculation, while a significant reduction in microbial ATP was recorded (47 %) after dual media filtration. After flocculation, microbial ATP measured using the ATP-direct method was lower than its concentration using the ATP-filtration method by approximately 20 %. However, the average difference in the four measured samples

65

across the pre-treatment train was less than 10 %. A higher number of samples was studied in the next sub-section to accurately assess the differences between the two methods.

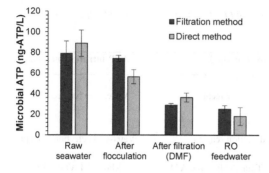

Figure 3.6: Comparison of microbial ATP concentration measured using the ATP-filtration method and the ATP-direct method for 4 samples collected during the pre-treatment of an SWRO desalination plant.

▪ Raw seawater

The microbial ATP concentrations of two ATP methods were compared in 125 raw seawater samples collected from the North Sea and the Tasman Sea. It was found that both methods report comparable microbial ATP concentrations ranging between 1 and 1,000 ng-ATP/L (Fig. 3.7). The measured microbial ATP concentration using the ATP-direct method is slightly lower (5 %) than the concentration based on the ATP-filtration method. An ANOVA analysis showed a significant correlation (R^2 = 0.95, p-value = $1.9 \times 10^{-80} \ll 0.001$, n = 125) between the two methods. The significance of the correlation between the two methods was tested using a "t-test of 2 samples assuming equal variances", which showed that the p-value (0.62) was much higher than alpha (α = 0.05) and t-Stat (0.5) was much lower than t critical two–tail (1.97), indicating that the correlation between the two methods is significant with 95 % confidence. These results clearly show that the measured microbial ATP concentrations by both the ATP-filtration and ATP-direct methods were comparable and correlated well.

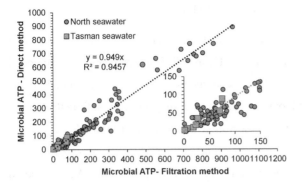

Figure 3.7: Correlation between the ATP-filtration and ATP-direct methods for 125 seawater samples collected from North Sea water, The Netherlands and Tasman Sea water, Australia.

3.3.6 Applications of the ATP-filtration method
  - **Monitoring microbial ATP in a full-scale SWRO plant**

The microbial ATP concentration was monitored along the treatment process of an SWRO desalination plant (with and without chlorination, 1 mg Cl_2/L in the intake). Samples were collected from raw seawater, after the first stage of dual media filtration (DMF1), after the second stage of dual media filtration (DMF2), after cartridge filtration, SWRO permeate and after remineralization.

Before intermittent chlorination, microbial ATP concentration in the influent was 525 ng-ATP/L, and a significant reduction of microbial ATP (> 95 %) was recorded through DMF1 incorporated with pre-coagulation (3.8 mg-$FeCl_3$/L), in which microbial ATP after DMF1 decreased to 20 ng-ATP/L (Fig. 3.8). Insignificant ATP removal was measured (< 1 %) through DMF2 and the cartridge filter. The remaining 4 % of the microbial ATP was removed through the SWRO membrane, in which microbial ATP concentrations after the 1st pass of the SWRO membrane and after remineralization were low (below 0.4 ng-ATP/L). The measured microbial ATP concentration in the RO permeate was 7 times higher than the LOD of the ATP-filtration method. The ATP-filtration method is versatile because it can be used to measure microbial ATP concentration in seawater as well as SWRO permeate (freshwater) down to 0.06 ng-ATP/L.

When intermittent chlorination was applied, microbial ATP in the influent decreased to 27 ng-ATP/L (Fig. 3.8) due to the addition of 1 mg-Cl_2/L in the intake pipe. Microbial ATP concentration after DMF1 was higher than the microbial ATP concentration in the influent, which could be due to biomass detachment during chlorination from the biofilm present in the media filter. The same observation was noted after DMF2 but at a lower magnitude, which may be because of the low concentration of free chlorine reaching the second stage of DMF or the low biofilm formation in DMF2. Microbial ATP decreased from 22 to 10 ng-ATP/L through 5 μm pore size cartridge filtration. Microbial ATP concentration after SWRO was also low (0.2 ng-ATP/L), which is 3 times higher than the LOD of the ATP-filtration method. However, a higher microbial ATP concentration was observed after remineralization compared to the measured concentration in the SWRO permeate. The high microbial ATP concentration after remineralization could indicate bacterial re-growth, which might occur due to nutrients originating from the added contaminated salts (calcium and fluoride).

This result clearly shows the effect of added chlorination in the intake on each process along the RO pre-treatment. It also demonstrates the applicability of the ATP-filtration method to monitor the biological conditions through the pre-treatment of SWRO and after RO membrane systems. Moreover, this result suggests that measuring microbial ATP could be used to monitor the biological treatment performance and might be useful for optimizing the treatment process.

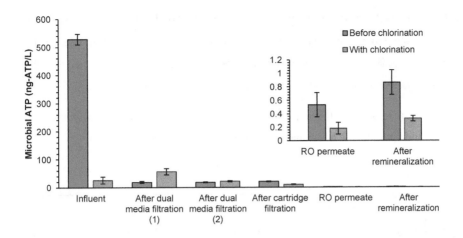

Figure 3.8: Monitored microbial ATP concentrations using the ATP-filtration method for 6 samples collected through the treatment processes of a full-scale SWRO desalination plant. All data are plotted as average ± standard deviation (n=3).

- **Monitoring of a media filtration**

The ATP-filtration method was used to monitor the performance of a seawater media filtration in terms of backwashability and microbial removal through the filter. For this purpose, microbial ATP was measured in raw North Sea water (influent) and in the filtrate of the media filtration (effluent) within one filtration cycle (48 h).

In the filtrate of the seawater media filtration, microbial ATP ranged between 92 and 140 ng-ATP/L before backwashing and ranged between 150 and 200 ng-ATP/L within the first hour after backwashing (Fig. 3.9). This result indicates that 7 minutes of backwashing could be enough to keep the media filter functional for longer time.

The removal of microbial ATP through the media filtration ranged between 65 to 85 %. The trend of microbial ATP concentrations of the filtered seawater followed the same trend of microbial ATP concentrations in raw seawater. Microbial ATP in the raw seawater increased from 1,050 to 3,000 ng-ATP/L over day 1 and decreased to 1,350 ng-ATP/L over day 2. Similarly, microbial ATP in the filtered seawater increased from 150 ng-ATP/L (after backwashing) to 450 ng-ATP/L over day 1 and decreased to 210 ng-

ATP/L over day 2. The dissimilar trend of microbial ATP of day 1 and day 2 is attributed to the seawater temperature, which increased from 12 to 23 °C on day 1 and declined to 16 °C on day 2. Higher bacterial growth and production are commonly reported at higher seawater temperature (White et al. 1991).

This result may suggest the applicability of using microbial ATP measurements to monitor and optimize the performance of seawater media filtration.

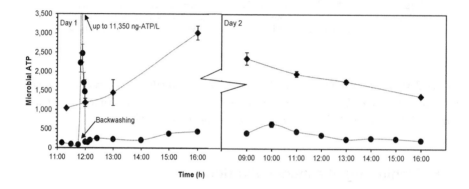

Figure 3.9: Monitored microbial ATP concentrations before and after seawater media filtration over one filtration cycle. Symbols: (♦) Raw seawater, (●) Filtered seawater.

- **Monitoring of bacterial growth in seawater**

The ATP-filtration method was applied to monitor the bacterial growth potential using the indigenous microbial consortium. Samples across the treatment line of a pilot-scale plant (namely, raw seawater, after media filtration and after ultrafiltration) were collected, pasteurized, inoculated with 10,000 intact cells/mL, incubated at 30 °C and monitored daily over 5 days based on microbial ATP.

The maximum microbial ATP concentration was reached after 2 days starting from 2 ± 1 ng-ATP/L (Fig. 3.10). The maximum microbial ATP reached depends on the nutrients available in the seawater. The maximum growth of raw seawater was 327 ng-ATP/L and decreased to 183 ng-ATP/L after the media filtration and to 160 ng-ATP/L after ultrafiltration. High reduction (44 %) of bacterial growth potential or nutrient removal was achieved by media filtration, while additional 7 % of bacterial growth potential reduction was achieved through ultrafiltration. The lower reduction of bacterial growth potential in the ultrafiltration might be because most of particulate organic matter was previously removed in the media filtration. This result is in line with the findings of Kim et al. (2011) who observed 38 % reduction of organic matter through seawater media filtration.

It can be seen from these results that the ATP-filtration method can be used for monitoring the bacterial growth in seawater.

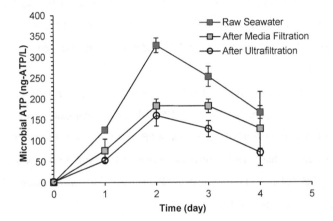

Figure 3.10: Monitored bacterial growth potential based on microbial ATP using the ATP-filtration method. Samples were collected from raw seawater, after media filtration and after ultrafiltration of a pilot-scale plant (The Netherlands). Data of Day 0 is the initial microbial ATP concentration after inoculation.

3.4 CONCLUSIONS

- A new method was developed to measure microbial ATP in seawater by incorporating a filtration step to concentrate the sample and to overcome the interference of salt. The measured microbial ATP concentration using the ATP-filtration method was comparable ($\pm$ 5 % difference, R^2 = 0.95, n = 125) to the concentration measured using the ATP-direct method in seawater.

- A very low limit of detection (0.06 ng-ATP/L, equivalent to 70 cells/mL) was obtained based on the ATP-filtration method, which is 3 times lower than the measured microbial ATP concentration in an SWRO permeate.

- Microbial ATP concentration in North Sea water samples correlated with intact cell concentration measured by flow cytometry (R^2 = 0.72, Rho = 0.88, P $\ll$ 0.001, n = 100). The average of the microbial ATP per marine bacterial cell was 8.59×10^{-7} ng-ATP/cell.

- The ATP-filtration method was applied to measure microbial ATP concentrations along the pre-treatment and permeate of an SWRO desalination plant. In the SWRO plant, significant reduction of microbial ATP (> 95 %) was recorded through the first stage of dual media filtration incorporated with inline pre-coagulation (3.8 mg-$FeCl_3$/L). A low microbial ATP concentration was measured in the SWRO permeate (0.2 ng-ATP/L).

- The ATP-filtration method was used to assess the microbial removal and backwashability of a seawater media filtration. The removal of microbial ATP through the media filtration ranged between 65 to 85 % and found that 7 minutes of backwash could be sufficient to keep the media filter functional.

- Furthermore, the ATP-filtration method was employed to monitor bacterial growth potential across the treatment line of a seawater pilot plant, in which the reduction in the bacterial growth potential was 44 and 7 % through the media filtration and ultrafiltration, respectively.

3.5 ACKNOWLEDGMENTS

We thank David Grasso for helping obtain access and collecting samples from the full-scale seawater desalination plant. Special thanks are due to Sophie Boettger, Almohanad Abusultan, Chidiebere Nnebuo and Maud Salvaresi for their assistance in the preliminary work of this research.

3.6 SUPPLEMENTARY DATA

S3.1 The retention of microorganisms on 0.1 µm filter

Table S3.1: The retention of microorganisms on 0.1 µm filter as intact cell concentration measured by flow cytometry.

	Intact cell concentration measured by flow cytometry (cell/mL)							
	Sample #1	Sample #2	Sample #3	Sample #4	Sample #5	Sample #6	Sample #7	Sample #8
Raw Seawater	862,800	529,600	616,607	390,247	334,740	733,393	657,460	2,114,973
Filtrate of 0.1 µm filter	140	250	40	220	107	100	153	47
LOD	500	500	500	500	500	500	500	500
Retention (%)	99.94	99.91	99.92	99.87	99.85	99.93	99.92	99.98

S3.2 The effect of salinity of rising solution on the captured marine microorganisms

To study the osmotic stress of captured marine microorganisms during rinsing, microbial ATP concentration of a seawater sample (Kamperland, The Netherlands) was measured using rinsing solutions (ASW$_{all\ ions}$) with different salt concentrations ranging between 1 and 60 g/L. It was found that marine microorganisms of North Sea water (living in a 32.5 g/L salt concentration) could survive with rinsing solutions ranging between 10 and 35 g/L.

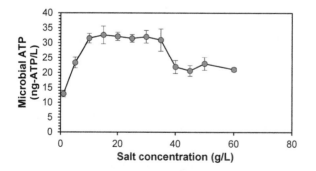

Figure S3.1: Effect of salt concentration on the survival of marine bacteria. Error bars are based on the standard deviation of triplicate measurement.

S3.3 Survival of bacteria in ASWall ions

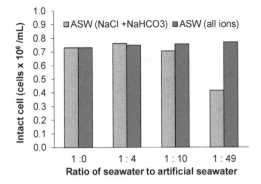

Figure S3.2: Intact cell concentration measured by flow cytometry of the seawater sample diluted with artificial seawater at different dilution ratios (1:4, 1:10, and 1:50). These ratios were compared to an undiluted seawater sample (1:0). Two different artificial seawaters were tested, i.e., ASW(NaCl+NaHCO3) and ASW (all ions). The concentration of intact cells was calculated back after dilution.

S3.4 The necessity of calcium or magnesium for the survival of marine bacteria

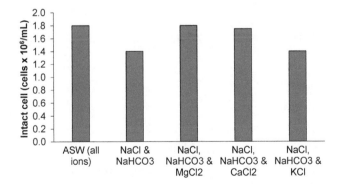

Figure S3.3: Intact cell concentration measured by the flow cytometry of indigenous marine bacteria in different artificial seawaters, including various compositions of water ions. This figure shows that the presence of calcium or magnesium is essential for the survival of marine bacteria.

S3.5 Relationship between the sample volume and the measured luminescence signal

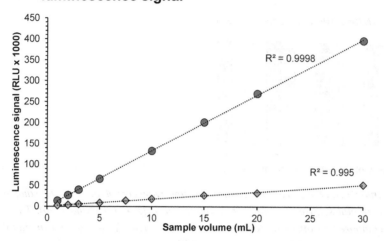

Figure 3.S4: The relationship between the filtered seawater sample volume (from 1 to 30 mL) and the measured luminescence signal of 2 samples. The figure shows that there is no effect of sample volume on determining the microbial ATP concentration for the tested range.

4

BIOFOULING POTENTIAL USING ATP-BASED BACTERIAL GROWTH POTENTIAL METHOD

Various bacterial growth potential (BGP) methods have been developed recently with the aim of monitoring biofouling in seawater reverse osmosis (SWRO) systems such as assimilable organic carbon and bacterial regrowth potential. However, the relationship between these methods and biofouling in full-scale SWRO desalination plants has not yet been demonstrated. In this chapter, a new method to measure bacterial growth potential (BGP) in seawater using indigenous microbial culture has been developed based on microbial ATP. Moreover, an attempt is made to investigate the correlation between BGP of SWRO feed water and the chemical cleaning frequency in three SWRO plants using the newly ATP-based BGP method.

The detection limit of BGP is investigated to 13 µg glucose/L. Low variations of bacterial yield were observed for five different seawater locations. BGP was monitored in raw seawater of North Sea and seasonal variations is observed ranging between 45 µg-glucose-C/L in the winter and 385 µg-glucose-C/L in the autumn.

The BGP method was applied to assess the pre-treatment performance of three full-scale SWRO plants with different pre-treatment processes. Dual media filtration (DMF) showed the highest BGP removal (> 50 %) in two SWRO plants. Removal of BGP and hydrophilic organic carbon in dissolved air floatation combined with ultrafiltration was similar to the removal achieved with DMF in combination with inline coagulation. For the three SWRO plants investigated, a higher BGP in the SWRO feed water corresponded to a higher chemical cleaning frequency. However, more data is required to confirm if a real correlation exists between BGP and biofouling in SWRO plants.

Keywords: Seawater reverse osmosis; Biofouling; Assimilable organic carbon; Bacterial growth potential; Adenosine triphosphate. Predicting biofouling in the feed of reverse.

This chapter has been published as **Almotasembellah Abushaban**, Sergio G. Salinas-Rodriguez, Nirajan Dhakal, Jan C. Schippers, Maria D. Kennedy (2019). Assessing pre-treatment and seawater reverse osmosis performance using an ATP-based bacterial growth potential method. *Desalination* V. 467 (2019) p210-218.

4.1 INTRODUCTION

Biofilm formation on reverse osmosis (RO) membrane surfaces is inevitable (Van Loosdrecht et al. 2012) and may cause biofouling in some cases. Biofouling occurs when biofilm formation is excessive to the extent that operational problems arise (Vrouwenvelder and Van der Kooij 2001). To monitor biofouling in full-scale RO plants, head loss is commonly monitored across the first stage of the RO. Once head loss increases to a significant level (usually about 15 % increase from the initial head loss), membrane cleaning-in-place (CIP) is applied to maintain the desired permeability. The frequency of cleaning primarily depends on the biofouling potential of the feed water and the operational conditions (flux, pressure, concentration polarization and CIP efficiency) of SWRO (Qiu and Davies 2015, Saeki et al. 2016).

An early warning system to predict biofouling potential seems more suitable than taking action after pressure drop/ head loss has increased (Vrouwenvelder et al. 2011, Jeong 2013). Early warning systems may allow optimization of RO pre-treatment processes. However, to date, there are no methods or tools available that can predict biofouling in membrane-based desalination systems. The membrane fouling simulator (MFS) and biofilm formation rate (BFR) can be used to monitor biofilm development on a membrane surface, but the biofilm formation in these systems/takes almost the same amount of time needed for biofilm formation on a RO membrane surface (Dixon et al. 2012).

Recently, the use of growth potential methods has gained high interest among researchers as they may be directly linked to biofilm formation on RO membrane (LeChevallier 1990, Jeong et al. 2013b, Wang et al. 2014). These methods include assimilable organic carbon (AOC) (Van der Kooij et al. 1982), bacterial regrowth potential (BRP) (Withers and Drikas 1998, Dixon et al. 2012), and biomass production potential (BPP) (Stanfield and Jago 1987, Van der Kooij and Van der Wielen 2013). The relationship between these methods and biofouling development in full-scale plants has not yet been determined. In fresh water, Hijnen et al. (2009a) reported that 1 µg/L of AOC (as acetate) added to MFS feed water led to a significant pressure drop within 3 months. Weinrich et al. (2016) tested the biofouling of 30 and 1,000 µg-C/L on a bench-scale SWRO membrane and reported higher fouling on the RO membrane surface with 1,000 µg-C/L (as acetate) than with 30 µg-C/L in RO feed.

The AOC method was initially developed for freshwater by Van der Kooij et al. (1982) and was measured by pasteurizing the sample (at 70 °C for 30 min), inoculating it with *Pseudomonas fluorescens* P17 bacteria, incubating it over time (for 2 weeks) and measuring bacterial growth using plate counting. In this method, one pure strain (*Pseudomonas fluorescens* P17) is used, which cannot completely assimilate AOC due to its lack of exo-enzymes and interactions between different bacteria. *Spirillum sp. NOX* (NOX) was later added together with P17 by Van der Kooij and Hijnen to utilize oxalic acid for bacterial growth (Van der Kooij and Hijnen 1984, Van der Kooij 1992). Although these two strains (P17 and NOX) utilize a wide range of easily biodegradable compounds, they cannot utilize more complex compounds such as polysaccharides and proteins. Sack et al. (2010) introduced an additional bacterial culture (*Flavobacterium johnsoniaestrain A3*) to the freshwater AOC test which target polysaccharides and proteins as nutrients for growth.

Another approach is the use of an indigenous microbial consortium to further broaden and diversify the substrate utilization range in comparison to a single pure culture. Ross et al. (2013) demonstrated that bacterial growth using an indigenous microbial consortium was higher (> 20 %) than bacterial growth of pure strains and provides a more realistic interpretation of growth potential in water. Several AOC methods have been developed using an indigenous microbial consortium for freshwater based on microbial adenosine triphosphate (ATP) (Stanfield and Jago 1987), turbidity (Werner and Hambsch 1986), and total cell counts (Hammes and Egli 2005).

Reporting growth potential-based methods as an AOC concentration is questionable as the calibration is performed using only one carbon source (glucose or acetate), while in real water, AOC is a mixture of different carbon sources. To overcome this problem, Withers and Drikas (1998) developed a turbidity-based BRP method to monitor bacterial growth in water distribution systems employing the typical procedure of the AOC method in which bacterial growth is reported as µg-C/L (acetate equivalent). Moreover, Van der Kooij and Van der Wielen (2013)developed the BPP method for drinking water in which the maximum bacterial growth and the cumulative biomass production are reported (in ng-ATP/L) without a conversion to carbon concentration (in μg-C/L).

In seawater, two AOC methods have been developed recently to measure the growth potential in the pre-treatment and in the feed of a SWRO membrane system by Weinrich et al. (2011) and Jeong et al. (2013b) using a single strain of bacteria (*Vibrio fischeri* and *Vibrio harveyi*, respectively). The use of a single bacterial strain allows normalization of the yield based on a carbon source, enabling conversion of bacterial growth to a carbon concentration. However, this method may not reflect the carbon utilization of indigenous microorganisms in seawater, and thus it may underestimate the nutrient concentration in seawater. In addition to the two AOC methods, Dixon et al. (2012) used a turbidity-based BRP method (developed by Withers and Drikas (1998)) to evaluate SWRO biofouling using an indigenous microbial consortium. Table 4.1 summarizes the available growth potential methods that can be applied in seawater.

The bacterial enumeration method employed to monitor growth potential depends on the bacterial culture. Conventional enumeration methods (i.e., heterotrophic plate counting, total direct cell count) are labourious, time consuming and limited to a small percentage of the overall bacterial count (Jannasch and Jones 1959). Weinrich et al. (2011) and Jeong et al. (2013b) used bioluminescence to monitor bacterial growth, as both methods employ luminescent bacteria (*Vibrio fischeri* and *Vibrio harveyi*, respectively). Due to the lack of fast and accurate bacterial enumeration methods, Dixon et al. (Dixon et al. 2012) and Quek et al. (2015) used turbidity and microbial electrolysis cell biosensor, respectively, to measure bacterial growth potential in seawater using an indigenous microbial consortium. Recently, new alternative methods that are fast, reliable, accurate and culture-independent have been developed in seawater with low level of detection, such as flow cytometry (FCM) (Van der Merwe et al. 2014, Dhakal 2017, Farhat et al. 2018) and ATP (Abushaban et al. 2018, Farhat et al. 2018, Abushaban et al. 2019b).

Table 4.1: Growth potential methods that can be applied in seawater (Weinrich et al. 2011, Dixon et al. 2012, Jeong et al. 2013b, Quek et al. 2015, Dhakal 2017)

Reference	Bacterial inactivation	Culture	Enumeration method	Expressed results
Weinrich et al. (2011)	Pasteurization (70 °C for 30 min)	*Vibrio fischeri*	Bioluminescence	µg-C as acetate equivalent
Dixon et al. (2012)	Filtration (0.2 µm)	Indigenous microorganisms	Turbidity	µg-C as acetate equivalent
Jeong et al. (2013)	Pasteurization (70 °C for 30 min)	*Vibrio harveyi*	Bioluminescence	µg-C as glucose equivalent
Quek et al. (2015)	-	Indigenous microorganisms	Microbial electrolysis cell biosensor	µg-C as acetate equivalent
Abushaban et al. (2018)	Pasteurization (70 °C for 30 min)	Indigenous microorganisms	Microbial ATP	µg-C as glucose equivalent
Dhakal et al. (2017)	Filtration (0.22 µm)	Indigenous microorganisms	Intact cell counts by FCM	µg-C as glucose equivalent
Farhat et al. (2018)	Filtration (0.2 µm)	Indigenous microorganisms	Total ATP and Total cell count by FCM	µg-C as acetate equivalent

The removal of bacterial growth potential along SWRO pre-treatment trains has been discussed in the recent literature using the newly developed methods. Weinrich et al. (2011) reported high variations (20 – 70 %) in AOC removal through a sand filter (Tampa Bay desalination plant) and 50 % AOC removal (from 20 to 10 µg C-acetate/L) in ultrafiltration (0.01 and 0.04 µm pore size) (Monterey Bay desalination plant). Moreover, Weinrich et al. (2015a) reported 43 % removal of AOC in the media filtration with inline coagulation (0.6 mg-Fe^{3+}/L) at the Al Zawarah desalination plant, UAE. This is similar to the reported removal by Abushaban et al. (2019b), in which 44 and 7 % removal of bacterial growth potential were observed through seawater glass media filtration (without coagulation) and ultrafiltration, respectively, in a pilot plant in the Netherlands. Whereas, Jeong et al. (2016) found insignificant AOC removal (4 %) through dual media filtration (DMF) combined with inline coagulation (ferric sulphate, dose is not mentioned) due to continues dosage of sodium hypochlorite to the seawater intake of Perth SWRO desalination plant. Weinrich et al. (2015a) studied the removal of AOC in three SWRO plants and reported higher AOC concentration in the SWRO feed due to chemical additions which may increase biofouling potential. The reported AOC in RO feed water ranged between 10 and 180 µg C/L as acetate. Thus, a preliminary AOC threshold of 50 µg C/L as acetate was suggested using growth kinetics and maximum yield of *Vibrio harveyi* bacteria in the saltwater applied in a pilot plant.

In this chapter, bacterial growth potential (BGP) is measured based on microbial ATP and using an indigenous microbial consortium. Using an indigenous microbial consortium and microbial ATP as an enumeration method may provide more accurate and representative information of bacterial growth in seawater. The BGP was monitored in raw seawater from the North Sea and measured along the pre-treatment of three full-scale SWRO desalination plants. Finally, an attempt was made to investigate if any correlation exists between BGP in SWRO feed water and the cleaning frequency (CIP) in SWRO plants based on three full-scale SWRO desalination plants.

4.2 MATERIAL AND METHODS

4.2.1 Sample collection and storage

Raw seawater samples were collected from the North Sea at the Jacobahaven pilot plant (Kamperland, The Netherlands) in sterile 500 mL amber-colour glass sampling bottles with tight-sealing screw caps. The samples were transported (90 km) to Delft (The Netherlands) in a cooler box (5 °C). The TOC and TCC-FCM of the collected seawater are 1.28 ± 0.85 mg/L and $0.9 \pm 0.28 \times 10^6$, respectively.

4.2.2 Preparation of artificial seawater

For the determination limit of detection (LOD) and bacterial yield of BGP method, artificial seawater (ASW) was used as a blank. ASW was prepared using Milli-Q water (Milli-Q® water Optimized purification, 18.2 MΩ.cm at 25°C, EC < 10 µS/cm, TOC < 30 µg/L, Millipore, USA) and analytical grade inorganic salts (Merck, USA). ASW was prepared to mimic similar ion concentrations to the average global seawater (Villacorte 2014) (23.668 g/L NaCl, 10.873 g/L $MgCl_2.6H_2O$, 3.993 g/L Na_2SO_4, 1.54 g/L $CaCl_2.2H_2O$, 0.739 g/L KCl, 0.213 g/L $NaHCO_3$, and 0.002 g/L Na_2CO_3). To eliminate the carbon contamination coming from salts, carbon was made volatile using high temperatures below the melting points of each salts. Therefore, sodium chloride (NaCl), sodium sulfate (Na_2SO_4), and potassium chloride (KCL) were heated at 550 °C in an oven furnace for 6 hours. Calcium chloride di-hydrate ($CaCl_2.2H_2O$) and magnesium chloride hexahydrate ($MgCl_2.6H_2O$) were heated at 100 °C for 20 minutes as their melting points are 176 °C and 117 °C, respectively. Sodium hydrogen carbonate ($NaHCO_3$) was not heated because its melting point is 50 °C. Salts were mixed (200 rpm) in Milli-Q water for 24 hours. The prepared ASW has the following properties (pH = 8.0 ± 0.1, EC = 52.6 ± 1.2 mS/cm, TOC < 100 µg/L and Total ATP < 0.05 ng-ATP/L).

4.2.3 Cleaning of glassware

All vials and caps were washed with a lab detergent (Alconox Ultrasonic Cleaner, Alconox, USA), rinsed with Milli-Q water (Milli-Q® water Optimized purification, 18.2 MΩ·cm at 25 °C, EC < 10 µS/cm, TOC < 30 µg/L, Millipore, USA) and submerged in 0.2 M HOCl (Merck, Millipore, USA) for 15 hours. Afterwards, they were rinsed again three times with Milli-Q water and were air dried. To eliminate potential organic contamination, the vials were heated in a furnace oven for six hours at 550 °C while the vial caps were bathed in a sodium persulfate solution (100 g/L, Merck, Millipore, USA) for 1 hour at 60 °C. Finally, the caps were rinsed with Milli-Q water and air dried.

4.2.4 Bacterial growth potential measurements

Measuring BGP in seawater comprises four steps, including bacterial inactivation, bacterial inoculation, incubation and bacterial enumeration (Fig. 4.1). Each step has been studied comprehensively (see supplementary data). Bacterial inactivation and inoculation were used as the microbial population during SWRO pre-treatment is not constant in terms of number and composition. Microbial inactivation allows the standardization of the initial microbial population by adding a constant inoculum concentration. Based on results shown in Section S4.2, both filtration and pasteurization can be used to inactivate the microbial population in seawater. However, due to the possibility of carbon release from virgin filters (Dhakal 2017), pasteurization was used. Moreover, sterilization was not used due to the possibility of carbon degradation at a high temperature (Section S4.2). The heating temperature during pasteurization was also tested, and it was found that there was no carbon degradation when seawater was heated at temperatures between 70 – 100 °C (Section S4.2). An inoculum concentration (100 – 20,000 cells/mL) was tested and

10,000 cells/mL was used to ensure sufficient cells for growth and to shorten the growth time to 2 days (Section S4.3), which agrees with the reported concentration in the literature (Hammes and Egli 2005). Negligible nutrient concentration (< 3 %) was estimated to be introduced into the seawater sample from the inoculum (Section S4.3.1). However, the incubation temperature has a significant effect on bacterial growth; the highest bacterial growth of indigenous microorganisms was achieved when the incubation temperature was similar to the original inoculum temperature (Section S4.4). This effect was overcome by using a calibration line for carbon and BGP at a constant incubation temperature for each seawater type. Using a calibration line and calculating bacterial yield allows the BGPs of different seawater samples at different locations to be compared.

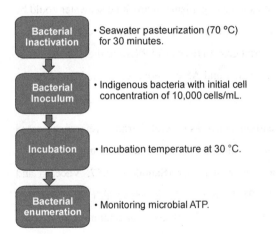

Figure 4.1: Procedure of measuring BGP in seawater based on microbial ATP.

4.2.5 Microbial ATP measurements in seawater

Microbial ATP was determined using the direct ATP method for seawater as described in. Abushaban et al. (2018). Briefly, total ATP and free ATP were measured to determine microbial ATP (microbial ATP = total ATP – free ATP). For the total ATP measurement, 100 µL of Water-Glo lysis reagent (Water-Glo kit, Promega Corp., USA) was added to 100 µL of the seawater sample in a 1.5 mL Eppendorf tube. The mixture (seawater and lysis reagent) and the Water-Glo detection reagent (Water-Glo kit, Promega Corp., USA) were heated at 38 °C for 4 min. An aliquot of 200 µL of the heated ATP detection reagent was added to the mixture. For the free ATP measurement, 200 µL of pre-heated (at 38 °C for 4 min) Water-Glo detection reagent was added to 100 µL of pre-heated seawater sample in a 1.5 mL Eppendorf tube. The bioluminescence signal was measured using a Promega GloMax®-20/20 luminometer. The measured bioluminescence signals were converted to the total ATP and free ATP concentrations based on 2 calibration curves.

4.2.6 Bacterial yield

To investigate the bacterial yield in North seawater, bacterial growth with different glucose concentrations (0, 10, 25, 50, 75 and 100 µg-C/L) was monitored (based on microbial ATP) in both real seawater (North Sea, The Netherlands) and artificial seawater (ASW) since the behaviour of indigenous microorganisms in artificial seawater could be different due to the presence of different substrate in real seawater. A correlation was established between the maximum bacterial growth (as ng-ATP/L) and the added glucose concentration. The bacterial yields in seawater and ASW were investigated based on the slope of the correlation line.

Glucose is used in this research as a carbon source as several literature references stated that glucose is a likely substance for assimilation in seawater and concentrations of 10^{-6} - 10^{-8} M glucose are known to be present in seawater (Saunders 1957, Vaccaro and Jannasch 1966, Chia and Warwick 1969). Moreover, Weinrich et al. (2011) reported higher bacterial growth of marine microorganisms with glucose concentration (0- 140 µg-C/L as glucose) than acetate.

4.2.7 The limit of detection of the ATP-based BGP method

The limit of detection (LOD) of the BGP method was determined using a microbial inoculum from the North Sea in 10 blanks in triplicate, in which ASW (TOC < 30 µg/L) was used as a blank. ASW was prepared as described in Abushaban et al. (2018). Nitrogen (20 µg-N/L as $NaNO_3$) and phosphorous (5 µg-P/L as $NaH_2PO_4.2H_2O$) were added to the blank to avoid bacterial growth inhibition due to Nitrogen and Phosphorous limitation. Bacterial growth was monitored based on microbial ATP (LOD = 0.3 ng-ATP/L) (Abushaban et al. 2018). The maximum bacterial growth within 5 days (14.7 ± 1.6 ng-

ATP/L) was used as BGP. LOD of BGP (19.5 ng-ATP/L,13 µg-C/L as glucose) was determined using the following equation (Taverniers et al. 2004).

$$LOD = Average\ of\ 10\ blanks + 3 \times standard\ deviation\ of\ 10\ blanks$$

4.2.8 Monitoring BGP of the North Sea

BGP, algal cell concentration and water temperature were monitored from the North Sea at the Jacobahaven pilot plant (Kamperland, The Netherlands) from January 2016 to January 2017. Raw seawater samples were collected weekly in sterile 500 mL amber-colour glass sampling bottles and transported (90 km) to Delft (The Netherlands) in a cooler box (5 °C). The summary of the properties of the collected samples is as follows: total organic carbon (TOC) = 1.28 ± 0.85 mg/L, total cell concentration measured by flow cytometry = $0.9 \pm 0.28 \times 10^6$ cells/mL, pH = 8.0 ± 0.1 and EC = 52.6 ± 1.2 mS/cm.

4.2.9 Organic carbon and biopolymer measurement

Liquid chromatography - Organic Carbon Detection was used to measure the hydrophilic organic carbon and biopolymer concentrations. The measurement and analysis of the samples were performed according to the protocol described by Huber et al. (2011). Seawater samples were shipped in a cooler box (5 °C) to Doc-labor Huber lab (Karlsruhe, Germany) for analysis.

4.2.10 Monitoring BGP along the pre-treatment of three SWRO plants

BGP was measured along the pre-treatment trains of three large (capacity > 120,000 m³/day) SWRO desalination plants located in the Middle East and Australia. The raw seawater of the three SWRO plants comes from open intakes, in which plant A and plant B have similar characteristics of raw seawater properties (Table 4.2). The SWRO pre-treatment of the three plants are different. Fig. 4.2 shows the treatment schemes and the locations of all collected samples. Brief specifications and operating conditions of the three plants are presented in Table 4.3.

Table 4.2: The properties of raw seawater of the three SWRO desalination plants.

	Plant A	Plant B	Plant C
Salinity (mS/cm)	69 - 71	69 - 71	54 - 60
TDS (g/L)	49 - 50	49 - 50	34 - 35
pH	8.3 - 8.6	8.3 - 8.6	8.1 - 8.3
Turbidity (NTU)	4 - 10	4 - 10	1 - 2
Water temperature (°C)	22 - 30	22 - 30	18 - 25
DOC (mg-C/L)	1.1 ± 0.1	1.1 ± 0.1	0.9 ± 0.1
Silt density index (%/min)	4.7 ± 0.4	4.7 ± 0.4	4.1 ± 0.3
MFI-UF (s/L$_2$)	2050	2050	2150
Chlorophyll a (µg/L)	0.6	0.6	NA
Algal concentration (cell/mL)	600	600	NA

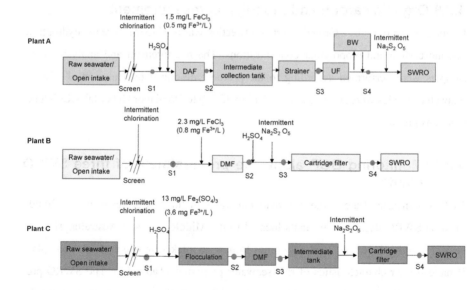

Figure 4.2: The treatment schemes of the three SWRO desalination plants in the Middle East and Australia, and the locations of collected samples for BGP monitoring.

Table 4.3: Operating conditions of the three SWRO desalination plants.

	Plant A	Plant B	Plant C
Pre-treatment	Coagulation + DAF + UF + cartridge filtration	Coagulation + Dual media filtration + cartridge filtration	Coagulation + flocculation + Dual media filtration + cartridge filtration
pH adjustment	at 7.9 in the intake by H_2SO_4	at 7.4in the SWRO feed by H_2SO_4	No adjustment
Coagulation dosage (mg-Fe^{3+}/L)	1.5 mg-$FeCl_3$/L (0.5 mg-Fe^{3+}/L)	2.3 mg-$FeCl_3$/L (0.8 mg-Fe^{3+}/L)	13 mg-$Fe_2(SO_4)_3$/L (3.6 mg-Fe^{3+}/L)
Type of filtration	Vertical ultrafiltration	Pressurised dual media filter	Gravity dual media filter
Type of media		Anthracite and sand	Coal and sand
Depth of media filter		1 m	1.6 m
Filtration cycle	1 h	24 – 48 h	48 h
Filtration rate (m/h)	0.06 (flux = 60 L/m^2/h)	11- 14	10 – 12
Estimated contact time	< 10 sec	4 - 5 min	8 - 9 min
Backwash protocol	Water	Air and water	Air and water
Antiscalant dosing	Yes	Yes	Yes
SWRO recovery	40 %	40 %	40 %

4.3   RESULTS AND DISCUSSION

4.3.1 Bacterial yield of indigenous microbial consortia

The conversion of microbial growth to carbon concentrations is only possible if the bacterial yield is known. For an indigenous microbial consortium, the bacterial yield needs to be determined for each location as it may vary depending on the microorganisms present in the inoculum (Weinrich et al. 2010, Wang et al. 2014). Bacterial yield can be investigated by determining the correlation between the carbon concentration and BGP for a specific location. Having this correlation also allows BGPs of different seawater samples of different locations to be compared.

Bacterial yields of the indigenous microbial consortium in seawater and ASW were investigated using glucose as a carbon source (Fig. 4.3). Good correlations in seawater ($R^2 = 0.98$) and ASW ($R^2 = 0.99$) were observed between BGP and the added glucose concentration. The higher intercept point of the seawater line (66.8 ng-ATP/L) compared with the ASW line (16 ng-ATP/L) is due to the presence of dissolved organic compounds in the seawater (natural background level). The slope of the correlation line in seawater (1.56 ng-ATP/ µg C-glucose) was slightly (9 %) higher than in ASW (1.43 ng-ATP/ µg C-glucose), revealing that the bacterial yield in seawater is greater. The difference in the bacterial yields could be attributed to the loss of some marine bacteria when they are placed in ASW, which is not their natural environment. The use of a different substrate in seawater may provide a higher bacterial yield (Vallino et al. 1996).

Similarly, bacterial yields of the Tasman Sea, Arabian Sea, Persian Gulf and Gulf of Oman were also determined (Table 4.4) using an indigenous microbial consortium collected on situ at each location. The bacterial yield ranged between 1 to 1.5 ng-ATP/µg C-glucose. The difference in the bacterial yield is attributed to several reasons, including the bacterial diversity present in the seawater and their activity, the carbon (as glucose) utilization rate and the seawater temperature.

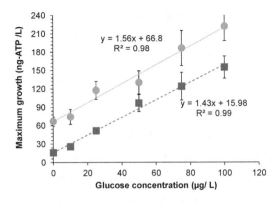

Figure 4.3: The correlation between added glucose concentration and the BGP in seawater (●) and artificial seawater (■).

Table 4.4: The bacterial yields of different microbial consortium of different seawaters.

Source of microbial consortium	Seawater temperature during sampling	Electrical Conductivity (mS/cm)	Bacterial yield (ng-ATP/ µg C-glucose)
North Sea	7 °C	52 - 54	1.5 ± 0.1
North Sea	20 °C	52 - 54	1.4 ± 0.1
Tasman Sea	25 °C	50 - 52	1.0 ± 0.1
Arabian Sea	22 °C	54 - 56	1.3 ± 0.2
Gulf of Oman	30 °C	55 - 56	1.2 ± 0.2
Persian Gulf	42 °C	69 - 71	1.3 ± 0.2

4.3.2 The limit of detection of the ATP-based BGP method

The average BGP of the blank after inoculation with marine microorganisms was 14.7 ± 1.6 ng-ATP/L. Thus, the LOD of the ATP-based BGP method was calculated to be 19.5 ng-ATP/L ($14.7 + 3 \times 1.6 = 19.5$). The bacterial growth in the blank indicates the presence of low concentrations of carbon, which could be introduced from several factors including the seawater inoculum (~5 ng-ATP/L), presence of nutrients in the (analytical grade) salts as well as the Milli-Q water used to make up ASW, and contamination from glassware and the surrounding environment. In this research, the blank was not subtracted from the measured BGP of seawater samples, as the origin of the nutrients in the blank is not known. Moreover, nutrient concentrations can vary in time as they originate from multiple sources as mentioned above.

Using the investigated bacterial yield of North Sea bacteria in seawater (1.56 ng-ATP/ μg C-glucose) and in ASW (1.43 ng-ATP/ μg C-glucose), the LOD of the BGP (19.5 ng-ATP/L) method was approximately 13 μg-C-glucose/L (19.5/1.5), respectively. Jeong et al. (2013b) reported 0.1 μg glucose-C/L of LOD in the AOC method in seawater using *Vibrio fischeri* bacteria. However, the reported LOD was calculated after subtracting the AOC of the blank, which was more than 50 μg C-glucose/L. To convert the LOD to C-acetate, Weinrich et al. (2011) found that the glucose utilization by *Vibrio harveyi* bacteria was higher than the acetate utilization at a concentration below 150 μg-C/L. Assuming the difference in carbon utilization applies to the indigenous microbial consortium as well, the LOD of the BGP method will be less than 10 μg acetate/L. This is similar to the reported LOD by Werner and Hambsch (1986) and Hammes et al. (2010a) in freshwater, using an indigenous microbial consortium based on turbidity and total cell counts measured by flow cytometry, respectively. Van der Kooij and Hijnen (1984) reported the lowest LOD (1 μg C-acetate/L) of AOC in freshwater, in which plating counts was used to monitor the growth of P17 and NOX.

Lowering the LOD to less than 5 μg C-glucose/L or even below 1 μg C-glucose/L would be ideal for measuring low BGP in the SWRO feed, particularly, in the winter. However, biofouling is not expected at low water temperatures with a low BGP. In this study, the lowest BGP measured in the SWRO feed was 70 μg C-glucose/L, in which the SWRO membrane was cleaned in place every 3 years (section 4.3.4) which was more than 5 times higher than LOD of the BGP method.

4.3.3 Monitoring of BGP in the North Sea

The BGP of raw North Seawater was monitored and a seasonal variation was observed ranging between 45 µg-C/L as glucose in the winter to 385 µg-C/L as glucose in the spring (Fig. 4.4). Two seasonal peaks of BGP were obtained in early spring (April) and in autumn (September/October). The BGP and algal cell concentration are similar to the observed trends in dissolved organic carbon (DOC) and chlorophyll a, respectively, by. Sintes et al. (2010) in the coastal North Seawater. They reported higher DOC values in the spring and autumn and lower DOC values in the winter and high chlorophyll a concentration in the spring.

Low algal concentration and BGP were observed at low water temperature (< 10 °C, November - February). In March, BGP and algal concentration increased indicating a spring algal bloom. A similar finding was observed by Huck et al. (1991), in which a higher AOC concentration was observed in the spring due to algae blooming. However, algal concentration further increased in April and May from 150 to 410 cells/mL while BGP decreased from 350 to 60 µg glucose-C/L. The decrease in BGP in the late spring could be attributed to the high nutrient utilization by algae during an algal bloom. Later, in the summer, despite the higher water temperature, algal concentration decreased to very low concentrations (50 cells/mL) while BGP increased to 300 µg glucose-C/L. The reduction in algal concentration in the summer could be due to the growth of other microorganisms that use algae as a source of nutrients (such as Daphnids and Rotifer) (Watanabe et al. 1955, Gilbert 1985). Thus, the BGP increased due to low algal concentration thus less competition for nutrients and/or due to released carbon from marine bacteria and algae (Bendtsen et al. 2002). The high BGP measured in autumn is consistent with the reported trend by Camper (2001), who monitored AOC in 64 surface water treatment plants. LeChevallier et al. (1996) monitored AOC and coliforms in 31 full-scale water plants and reported the same trend.

The correlation between BGP, algal cell concentration and water temperature was not evident all the year because both water temperature and the presence of algae influence BGP. For instance, BGP may only correlate with algal concentration during specific seasons (i.e. algal bloom in March/April) since very low algal concentrations were observed during the rest of the year. Moreover, a correlation might be possible between BGP and water temperature when algae does not play a role.

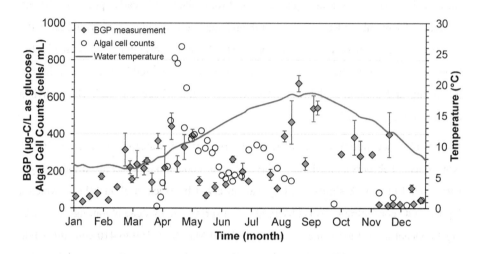

Figure 4.4: BGP, algal cell concentration and water temperature throughout 2016 in the North Sea raw seawater at the Jacobahaven pilot plant (Netherlands).

4.3.4 Monitoring of BGP in three full-scale SWRO plants

• Plant A

The SWRO pre-treatment of plant A consists of dissolved air flotation (DAF) and ultrafiltration (UF). The measured BGP of the raw seawater (before DAF) was 400 μg-C/L as glucose and decreased by 17.5 % to 330 μg-C/L as glucose after the DAF (Fig. 4.5a). The organic matter removal through the DAF is lower than that reported in literature. Shutova et al. (2016) reported 84, 25 and 16 % removal of biopolymers, low molecular weight acids (LMW-A) and DOC, respectively, in a lab scale DAF system fed with Gold Coast seawater with coagulant dose of 3 mg-Fe^{3+}/L (at pH 7.5). Whereas, the removal of biopolymers, LMW-A and DOC in the DAF system of plant A was 8, 2 and 2.5 %, respectively, (Table 4.5) using 0.5 mg-Fe^{3+}/L coagulant dose at pH 7.9. The low reduction of BGP through DAF could be attributed to the low coagulant dose (0.5 mg-Fe^{3+}/L), particularly, at high pH (pH 7.9). It has been reported by Shutova et al. (2016)

that coagulant dose in seawater DAF depends on pH, in which the optimal coagulation condition for organic matter removal is at low pH. The optimal coagulant dosage is 0.5 – 4 mg-Fe^{3+}//L at pH 5.5 and 4 – 12 mg-Fe^{3+}//L at pH 7.5 (Shutova et al. 2016).

A further removal of BGP (32.5%) was observed, mainly in the ultrafiltration (UF) system, where the BGP decreased to 200 µg glucose-C/L. This is consistent with the reported removal in the seawater UF. Weinrich et al. (2011) reported 50 % removal of the AOC concentration (from 20 to 10 µg C-acetate/L) through the ultrafiltration of the Moss Landing desalination pilot plant in California. Furthermore, Mathias et al. (2016) reported much lower dissolved organic matter removal (20 and 13 %) in 50 and 200 kDa seawater lab-scale UF membranes, respectively. The variation in the reported removal of organics depends on the type of natural organic matter (NOM) present in the seawater (Aoustin et al. 2001). It can be observed that the SWRO feed of plant A still supports a significant bacterial growth (> 200 µg glucose-C/L) despite DAF and UF being used as a pre-treatment. The total removal of BGP through the pre-treatment of plant A was 50 %.

- **Plant B**

The pre-treatment of plant B consists of single stage pressurized dual media filtration (DMF) after inline coagulation (0.8 mg Fe^{3+}/L). The measured BGP of the seawater before DMF was 350 µg glucose-C/L which decreased to 160 µg glucose-C/L after DMF (Fig. 4.5b). The significant reduction (55 %) of BGP through the DMF incorporation with inline coagulation could be attributed to the high biodegradation rate in the DMF. Similar findings were observed by Weinrich et al. (2011), in which the AOC removal through the sand filtration of a Tampa Bay desalination plant ranged between 23 and 80 %. BGP after the cartridge filtration (approx. 125 µg glucose-C/L) was similar to the measured BGP after DMF. The overall removal of BGP through the pre-treatment processes of plant B was 55 %, which was mainly due to carbon biodegradation in the DMF.

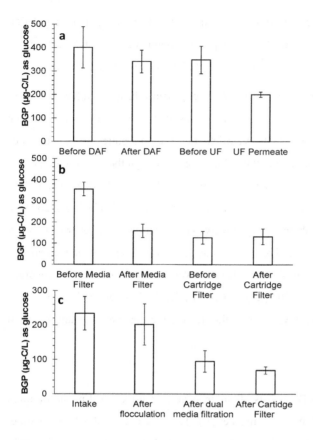

Figure 4.5: Monitored BGP along the RO pre-treatment trains of three SWRO desalination plants for 3 days (a) plant A in the Middle East, (b) plant B in the Middle East and (c) plant C in Australia.

- ## Plant C

The pre-treatment of plant C is a typical conventional treatment (coagulation, flocculation and gravity media filtration). BGP in the seawater intake was approximately 230 µg glucose-C/L (Fig. 4.5c), which is the lowest BGP in raw seawater among the three plants (plants A, B and C). Slight removal of BGP (15 %) was observed through the flocculation process due to the addition of coagulation with 13 mg/L of $Fe_2(SO_4)_3$, equivalent to 3.6 mg Fe^{3+}/L). Conversely, a significant removal of BGP (53 %) was noted in the DMF. The BGP removal of the conventional pre-treatment (coagulation, flocculation and gravity media filtration) of plant C (68 %) was higher than the observed BGP removal of the DMF incorporation with inline coagulation of plant B (55 %). The higher BGP removal of the conventional pre-treatment of plant C could be attributed to the longer contact time in the gravity DMF compared to the pressurized DMF of plant B and/or due to the higher coagulation dosage applied in plant C (3.6 mg Fe^{3+}/L). An insignificant BGP removal (4 %) through the cartridge filtration of plant C was found, as expected. The overall BGP removal in plant C was 72 %.

4.3.5 Comparing the removal of organic in the three SWRO plants

Comparing the overall removal of BGP and hydrophilic organic carbon through the pre-treatment of plants A and B shows that the combination of DAF and UF could provide approximately the same removal as that of the inline coagulation and DMF (Table 4.5). However, the UF (plant A) showed higher removal of the biopolymer fraction compared to the media filtration of plant B. Poussade et al. (2017) compared the removal of UF and media filtration and concluded that the removal rate of dissolved organic matter (expressed as UV254 absorbance and TOC) by media filtration was slightly better than that of UF, which was also found in the three SWRO plants studied here (based on BGP and hydrophilic organic carbon). The higher removal in media filtration compared to UF could be attributed to the biodegradation in the media filter as the contact time in media filter (4 - 5 min) is much longer than the contact time in UF (< 10 sec). Kim et al. (2011) tested the combination of DAF with DMF and found that DAF did not significantly improve the organic removal of DMF. This is also in agreement with the low removal of BGP, hydrophilic organic carbon and biopolymer observed through DAF in plant A. It should be noted that low coagulant dosage was added before DAF (plant A) and before DMF (plant B).

The BGP removal through conventional pre-treatment (plant C) was comparable to the removal achieved in DAF combined with UF (plant A) and inline coagulation incorporated with DMF (plant B). The overall BGP removal through the conventional pre-treatment was highest (72 %); however, the overall magnitude of the BGP removal (160 µg glucose-C/L) was lower than the removal in the other plants (Table 4.5). This is mainly because the raw seawater of plant C has a better quality than plants A and B. It should be noted that the coagulant dosage in plant C is very high compared to the applied coagulant dosage in plant B.

By comparing BGP's in the SWRO feed of three desalination plants, it can be seen that plant A has the highest BGP in the feed, while plant C has the lowest (Table 4.6). This finding indicates that the biofouling potential of plant A is the highest among the three SWRO desalination plants.

Table 4.5: Comparing the pre-treatment and their removal in the three studied plants.

	Plant A		Plant B	Plant C
	Inline coag. & DAF	Ultra-filtration	Inline coag. & pressurized DMF	Coag., Floc. & gravity DMF
Coagulation (mg-Fe^{3+}/L)	0.5	-	0.8	3.6
Contact time	NA	< 10 sec	4 - 5 min	8 - 9 min
BGP removal (µg-C/L)	70	130	190	156
BGP removal (%)	17 %	33 %	55 %	68 %
CDOC removal (µg/L)	27	133	151	NA
CDOC removal (%)	2.5 %	12 %	15 %	NA
Biopolymers removal (µg/L)	13	78	35	NA
Biopolymers removal (%)	8 %	46 %	29 %	NA
Humic subs. removal (µg/L)	10	28	59	NA
Humic subs. removal (%)	3 %	7 %	14 %	NA
LMW-N removal (µg/L)	11	1	15	NA
LMW-N removal (%)	6.5 %	< 1 %	9.5 %	NA
LMW-A removal (µg/L)	2	2	10	NA
LMW-A removal (%)	2 %	2 %	10 %	NA

NA: Not available

Table 4.6 Comparing the cleaning frequency and the BGPs of raw seawater and the RO feed of the three SWRO desalination plants.

	Plant A	Plant B	Plant C
BGP of raw seawater, µg glucose-C/L	400	350	230
BGP of RO feed, µg glucose-C/L)	200	128	70
Overall BGP removal, µg glucose-C/L (%)	200 (50 %)	222 (55 %)	160 (72 %)
CIP frequency (CIP's / year)	6	1	0.3

Investigating if a correlation exists between BGP in SWRO feed and biofouling in SWRO systems is complicated by several factors. Firstly, several types of fouling (scaling, particulate and organic/biofouling) may occur simultaneously in a SWRO plant. Secondly, to establish a correlation, a large number of SWRO desalination plants in different parts of the world need to be monitored for longer periods of time with different operating conditions. Thirdly, the widespread intermittent use of non-oxidizing biocides to combat biofouling in full-scale SWRO facilities makes establishing any real correlation between the BGP of SWRO feed water and the CIP frequency very difficult.

Despite these limitations, an attempt was made to investigate if a correlation exists between the measured BGP in SWRO feed water and the CIP frequency in the three SWRO plants. The CIP frequency (CIPs per year) was used as a surrogate parameter for biofouling, assuming that scaling and particulate fouling do not occur. This assumption is somehow justified as antiscalant is dosed prior to the SWRO membranes and thus should eliminate the occurrence of any scale. Furthermore, the SDI was always below (SDI < 3) in the SWRO feed water suggesting that particulate fouling was not significant in the SWRO plants studied.

From Figure 4.6 and Table 4.6, it can be observed that a higher CIP frequency corresponded to a higher BGP of SWRO feed water, suggesting that the BGP method is a promising indicator of biofouling potential in SWRO feed water. However, to establish a real correlation, more data needs to be collected and many more SWRO plants need to be monitored for longer periods of time with different operating conditions. Moreover, the monitoring program should be expanded to include a wide variety of seawater locations and pre-treatment technologies.

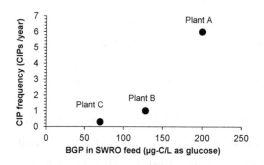

Figure 4.6: The relationship between the measured BGP in the SWRO feed and the cleaning frequency of the SWRO membrane of 3 desalination plants.

4.4 CONCLUSIONS

- A method based on microbial ATP was developed to measure BGP using an indigenous microbial consortium in seawater. The bacterial yield was measured in 5 locations and ranged between 1 and 1.5 ng-ATP/ μg C-glucose, thus indicating low variations of the bacterial yield of indigenous microorganisms in terms of microbial ATP. The limit of detection of the BGP method is 13 μg C-glucose/L.

- BGP of North Sea raw seawater was monitored over a period of 12 months, in which a seasonal variation was observed between 45 μg C-glucose/L in the winter and 385 μg C-glucose/L in the spring.

- The new method was applied to The maximum bacterial growth was reached 2 days faster using microbial ATP than its growth using intact cell concentration measured by flow cytometry.

- The use of 0.1 μm filtration and pasteurization for 30 minutes with and without pre-filtration (0.1 μm) were comparable to inactivate marine microorganisms in seawater.

- BGP was constant using 1,000 to 20,000 cells/mL inoculum concentration of indigenous microbial consortium. However, the time required for the microbial inoculum to reach the maximum production was different.

- Incubation temperature of indigenous bacteria had a significant effect on the BGP, in which the highest BGP was found at similar or close temperature to their original seawater temperature. However, a calibration line between carbon and BGP at constant temperature can be used to normalize for temperature.

- The method was applied to monitor BGP through the pre-treatment trains of three SWRO desalination plants with different pre-treatment processes. DMF showed the highest BGP removal (>than 50 %) in two SWRO desalination plants and this was attributed to the longer contact time in DMF filters (6 min) compared with UF (< 10 sec). The removal of DAF combined with UF was comparable to the removal of DMF in combination with inline coagulation (0.8 mg Fe^{3+}/L).

- A higher CIP frequency of the SWRO's corresponded to a higher BGP in SWRO feed water, suggesting that the BGP method is a promising indicator of biofouling potential in SWRO feed water. However, to establish a real correlation, more data needs to be collected and many more SWRO plants need to be monitored for longer periods of time, and the monitoring program should be expanded to include a wide variety of seawater locations and pre-treatment technologies.

4.5 ACKNOWLEDGEMENT

We thank Promega (Madison, USA) for providing ATP Water-Glo reagents and financially supporting this research. Special thanks are due to Nasir Mangal and Chidiebere Nnebuo for their assistance in the preliminary work of this research.

4.6 SUPPLEMENTARY MATERIAL

S4.1 Microbial enumeration method

The use of microbial ATP, as an indicator of microbial activity, was compared to the use of intact cell concentration measured by flow cytometry (ICC-FCM), as an indicator of cell counting. The comparison was made by monitoring the bacterial growth of indigenous bacterial consortium in raw seawater at the presence of different carbon concentration (0 - 1,000 µg C/L) as shown in Figure S4.1. The monitored bacterial growth based on microbial ATP showed that the initial microbial ATP concentrations (~ 4 ng/L) were rapidly grown, within 2 days, to its maximum growth with an average growth rate of 0.094 ±0.02 1/h (Fig. S4.1a). Afterwards, microbial ATP concentrations were dramatically decreased (day 3 – 7) which may indicate that most of added carbon (easy biodegradable) was consumed by the grown marine bacteria which in turn refer to the bacterial decay phase due to limited nutrients. Higher bacterial growth was observed at the higher added carbon concentration. It should be noted that the bacterial growth observed without spiking glucose (0 µg C/L) is due to the present nutrients in the raw seawater. On the other hand, the monitored ICC-FCM showed that the cell counts significantly increased within 2 days and thereafter it gradually increased further to the maximum growth on day 3 - 4 (Fig. S4.1b) with an average growth rate of 0.054 ± 0.01 1/h.

Comparing the calculated growth rate, higher growth rate based on microbial ATP (0.094 ±0.02 h^{-1}) comparing to the growth rate based on ICC-FCM (0.054 ± 0.01 h^{-1}). This is mainly due to the time required for the indigenous microbial inoculum to reach the maximum growth based on ATP (2 days) and ICC-FCM (4 days). Similar results were reported in freshwater; LeChevallier et al. (1993) measured AOC in freshwater based on ATP and reported that the time required for P17 and NOX bacteria to reach the maximum growth were 2 and 3 days at 30 °C and 15 °C, respectively, while Villacorte et al. (2017) used ICC-FCM for monitoring the bacterial growth of natural bacterial consortium in seawater and found a similar trend of bacterial growth, in which the maximum growth was observed after 4 days.

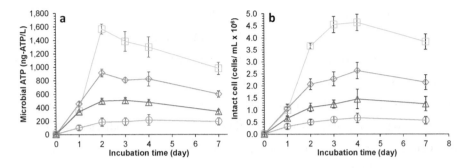

Figure S4.1: Monitoring of bacterial growth in raw seawater (North Sea, Kamperland, the Netherlands) spiked with different glucose concentrations based on (a) microbial ATP measurement and (b) intact cell concentration measured by flow cytometer. Symbols: (○) 0 μg-C /L, (Δ) 250 μg-C /L, (◊) 500 μg-C /L and (□) 1,000 μg-C /L.

Having the maximum bacterial growth based on microbial ATP faster than ICC-FCM indicates that the bacteria were very active during the growth phase (exponential phase) to an extent that the maximum activity (maximum microbial ATP) was observed 2 day before reaching to the maximum bacterial cell number (based on ICC-FCM). Another possible reason of the faster growth based on microbial ATP is that bacterial activity could vary from cell to another cell, in particular, when indigenous consortium is used. The variation of bacterial activity is taken into consideration in microbial ATP measurement while it is not the case in cell counts methods. This is a plus point of microbial ATP over cell counts.

For aforementioned reasons, it was decided to use microbial ATP as a microbial enumeration method in this research because (I) the time required for the maximum growth is shorter which allows the determination of BGP (as an indicator of SWRO biofouling) at early stage and (II) microbial ATP takes the variation of bacterial activity into consideration which is important when indigenous microbial culture is used.

S4.2 Microbial inactivation processes

Different microbial inactivation processes were tested including filtration (0.1 μm), pasteurization (70 °C), sterilization (121 °C), and filtration (0.1 μm) prior to pasteurization and sterilization by measuring the BGP of North Sea water sample. The

BGP of filtration, and pasteurisation with and without pre-filtration were comparable (198, 203, and 204 ng-ATP/L, respectively) as shown in Fig. S4.2a. Higher BGP values (301 and 305 ng-ATP /L) were observed when sterilization was applied with and without pre-filtration (respectively) which could be due to carbon degradation of natural organic matter (NOM) present in the seawater at high temperature (121 °C). The typical BGPs of pasteurization with and without pre-filtration and sterilization with and without pre-filtration reveal that pre-filtration of the seawater does not affect the measured BGP values. Li et al. (2017a) reported a similar finding in reclaimed water using 0.2 µm where the maximum bacterial growth using filtration with and without pasteurization was similar.

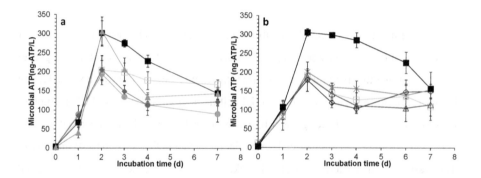

Figure S4.2: Comparing BGP of raw seawater sample (North Sea water) after applying (a) different inactivation processes and (b) different autoclaving temperatures. Symbols: (●) Filtration, (♦) Filtration and pasteurization at 70 °C, (▲) Filtration and sterilization at 121 °C, (□) Pasteurization at 70 °C, (◊) 80 °C, (△) 90 °C, (x) 100 °C and (■) Sterilization at 121 °C.

The BGP of different inactivation/autoclaving temperatures at 70, 80, 90, 100, and 121 °C was tested and found that no significant differences in the BGP of 70 – 100 °C (Fig. S4.2b). The BGP of 121 °C was 1.5 higher than the tested BGP at 70 to 100 °C which may indicate that the use of 70 – 100 °C can be applicable for microbial inactivation in seawater, while temperatures > 100 °C could degrade carbon present in the seawater to easier biodegradable carbon.

S4.3 Inoculum concentration

The effect of inoculum concentration from indigenous microbial consortium on BGP measurements were investigated at initial concentration ranging between 100 and 20,000 cells/mL as intact cells. No significant difference in BGPs of 1,000, 5,000, 10,000, and 20,000 cells/mL was observed (Fig. S4.3), in which the BGPs were 168, 173, 176, and 178 ng-ATP/L, respectively. However, the time required for maximum bacterial growth was different. The higher the inoculum concentration, the shorter time is needed for maximum growth. The time required of 1,000, 5,000, 10,000, and 20,000 cells/mL of inoculum was 4, 3, 2, and 2 days, respectively. Therefore, 10,000 cells/mL of inoculum concentration was used in order to shorten the time of growth phase. Similar finding was reported by Van der Kooij et al. (1982) and LeChevallier et al. (1993) in freshwater and Li et al. (2017a) in reclaimed water. The BGP of 100 cells/mL inoculum concentration (130 ng-ATP/L) was lower than the BGP of other higher inoculum concentrations suggesting that the lower inoculum concentration of indigenous microbial consortium cannot support the full growth. The lower BGP of 100 cells/mL could be due to lower diversity of indigenous microorganisms.

A slight difference between the BGP of 1,000, 5,000, 10,000 and 20,000 cells/mL of inoculum concentrations was noted from 168 to 178 ng-ATP/L due to introduced nutrients with the inoculum as the inoculum volume of untreated seawater (including nutrients) increased at high inoculum concentration. To estimate the added nutrients to the sample in this case, it was found that the variation in BGP is 10 ng-ATP/L (from 168 to 178 ng-ATP/L) for inoculum concentrations from 1,000 to 20,000 cells/mL. Thus, in average, additional 5 ng-ATP/L is expected when 10,000 cells/mL of inoculum concentrations is used, which is lower than 3 % in this case. More generally, assuming the BGP in raw seawater is 500 µg glucose-C/L (see section 2.3.7 and 2.3.8) and the volume of the inoculum is approximately 1 % of the sample (assuming intact cell concentration in the raw seawater is 1 million cells/mL). Thus, the introduced nutrients to the sample is 1 % of 500 ng-ATP/L which is 5 ng-ATP/L. This concentration is very low comparing to what have been measured in the SWRO feed (section 2.3.8). The inoculum can be treated to remove all nutrients in seawater by suspending the marine microorganisms in a saline buffer solution or artificial seawater, however, some marine microorganism could be lost during treatment.

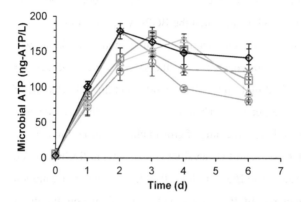

Figure S4.3: Bacterial growth curves of indigenous microbial consortium at different inoculum concentrations in seawater sample collected from the North Sea (The Netherlands). Symbols: (○) 100 cells/mL, (△) 1,000 cells/mL, (□) 5,000 cells/mL, (x) 10,000 cells/mL, and (◇) 20,000 cells/mL.

S4.4 Incubation temperature of Indigenous microorganisms

Incubation temperatures (from 5 °C to 37 °C) of two seawater samples from Arabian Sea and North Sea were tested with natural temperatures (during sampling) at 22 °C and 7 °C, respectively. It was observed that incubation temperature has a significant effect on the BGP (Fig. S4.4). In the Arabian seawater sample, the maximum BGP was achieved at 20 °C (similar to the natural water temperature during sampling) with 450 ng-ATP/L and then followed at 25, 30 and 37 °C with approximately 300 ng-ATP/L (Fig. S4.4a). On the contrary, the tested BGP of indigenous microorganisms in North seawater showed higher values at lower temperature (Fig. S4.4b), in which the highest BGP (980 ng-ATP/L) was achieved at 5 °C (similar to natural water temperature during sampling).

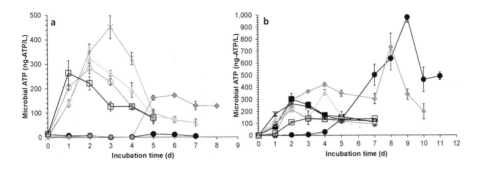

Figure S4.4: Bacterial growth of indigenous microbial consortium at different incubation temperatures of seawater samples collected from (a) Arabian Sea (Salalah, Oman) at 22 °C and (b) North Sea (The Netherlands) at 7 °. Symbols: (●) 5 °C, (♦) 10 °C, (■) 15 °C, (×) 20 °C, (▲) 25 °C, (○) 30 °C and (□) 37 °C.

Having the maximum BGP of Arabian seawater and North seawater at different incubation temperatures (at 20 and 5 °C, respectively) indicates that the bacteria in the used inoculums of Arabian Sea North Sea have been adapted to a seawater temperature close to their natural temperature. To overcome the variation of bacterial growth, calibration curve between carbon and bacterial growth at constant incubation temperature could be used to normalize for temperature. Moreover, having calibration curve allows comparing the BGPs of different seawater samples of different locations. Figure S4.5 shows the comparison of 2 seawater samples measured based on a calibration curve prepared at 20 °C and 30 °C incubation temperature, in which both maximum bacterial growth (as ng-ATP/L) provided the same BGP, since the sample and the calibration curve is prepared at the same incubation temperature.

Comparing BGP of Arabian Sea and North Sea is only possible when calibration curves with carbon are prepared for both seawater because both of them has different bacterial yield. Figure S4.4 shows that North Sea has higher growth (1,000 ng-ATP/L at 5 °C) than Arabian Sea (450 ng-ATP/L at 20 °C). However, when the BGPs of the 2 locations are compared based on a calibration line prepared at 30 °C incubation time, the BGP of Arabian Sea (215 µg C-glucose) is much higher than the BGP of North Sea (145 µg C-glucose) (Table S4.1).

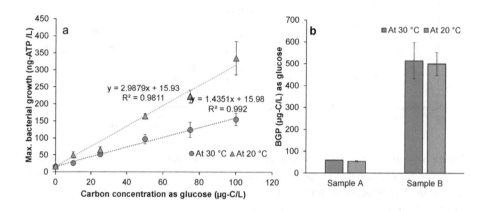

Figure S4.5: Comparison of 2 seawater samples from North Sea based on calibration curves prepared at 20 °C and 30 °C.

Table S4.1: Comparing BGP of Arabian Sea and North Sea with and without calibration curve with glucose.

	Arabian Sea	North Sea
Seawater temperature during sampling	22 °C	7 °C
Incubation temperature	30 °C	30 °C
BGP (ng-ATP/L)	280	220
Bacterial yield (ng-ATP/ µg C-glucose)	1.3	1.5
BGP (µg C-glucose)	215	145

5

CORRELATING BACTERIAL GROWTH POTENTIAL MEASUREMENT TO REAL TIME FOULING DEVELOPMENT IN FULL-SCALE SWRO

Recently, several methods were developed to monitor biological/organic fouling potential in seawater reverse osmosis (SWRO) systems such as bacterial growth potential (BGP). The correlation between these methods and biofouling in SWRO systems has not been demonstrated yet. In this research, the relation between BGP in SWRO feed water and SWRO membrane performance was investigated. For this purpose, the pre-treatment of a full-scale SWRO plant including dissolved air flotation (DAF) and two stage dual media filtration (DMF) was monitored for 5 months using different indices for particulate (silt density index and modified fouling index) and biological/organic fouling (BGP, orthophosphate, chromatography organic carbon).

Results showed that particulate fouling potential was well controlled through the SWRO pre-treatment as the measured indices in the SWRO feed water were below the recommended values. DAF in combination with coagulation (1 - 5 mg-Fe+3/L) achieved 70 % removal of orthophosphate, 50 % removal of BGP, 25 % removal of biopolymers, and 10 % removal of humic substances. Higher BGP in the SWRO feed water corresponded to a higher normalized pressure drop in the SWRO, suggesting the applicability of using BGP of SWRO feed water as a biofouling indicator in SWRO systems. However, to validate this conclusion, more SWRO plants with different pre-treatment systems need to be monitored.

Keywords: Desalination; Biofouling; fouling potential; seawater reverse osmosis; pre-treatment.

This chapter has been submitted to *Desalination* as **Almotasembellah Abushaban**, Sergio G. Salinas-Rodriguez, Moses Kapala, Delia Pastorelli, Jan C. Schippers, Subhanjan Mondal, Said Goueli, Maria D. Kennedy (2019). Correlating bacterial growth potential measurement to real time fouling development in full-scale SWRO.

5.1 INTRODUCTION

Membrane fouling is the main challenge that faces the operation of seawater reverse osmosis (SWRO) systems (Matin et al. 2011, Goh et al. 2018). Pre-treatment is commonly applied to improve water quality prior to reverse osmosis (RO), and thus to minimize/mitigate fouling issue in SWRO systems (Dietz and Kulinkina 2009, Henthorne and Boysen 2015). Almost all SWRO desalination plants require pre-treatment and the type of pre-treatment depends on the fouling potential of the raw seawater. Particulate fouling potential is commonly monitored by measuring the silt density index (SDI) and modified fouling index (MFI). Both SDI and $MFI_{0.45}$ are ASTM methods (ASTM D4189-14 2014, ASTM D8002 - 15 2015), in which MFI takes into account the occurrence of cake filtration (Schippers and Verdouw 1980). It has been reported that the maximum SDI_{15} (SDI of 15 min) value for acceptable SWRO feed water is 3 %/min (Badruzzaman et al. 2019).

However, to date, no standard method is available to monitor biological and organic fouling potential in SWRO systems. Monitoring biological and organic fouling potential through SWRO pre-treatment is important to improve SWRO performance (Jeong et al. 2016). For this reason, several methods are being developed and tested in SWRO desalination plants such as assimilable organic carbon (AOC) (Jeong et al. 2013b, Weinrich et al. 2013), bacterial regrowth potential (BRP) (Dixon et al. 2012) and bacterial growth potential (BGP) (Abushaban et al. 2019a).

The correlation between AOC and other biological/ organic and particulate fouling potential methods has been studied. Jeong and Vigneswaran (2015) found excellent correlations between AOC concentration and low molecular weight neutral (LMW-N) organics concentration ($R^2= 0.98$) and between AOC and the standard blocking index calculated from $MFI-UF_{10\ KDa}$ ($R^2= 0.97$). They suggested that $MFI-UF_{10\ KDa}$ can be used as a preliminary indicator of AOC and LMW-N. Weinrich et al. (2015b) observed that AOC concentration neither correlated with total organic carbon (TOC) nor UV_{254} in three full-scale SWRO desalination plants.

Investigating the correlation between biological/organic fouling indicators in SWRO feed water and real time biofouling development in SWRO membrane systems is complicated by a few factors (Abushaban et al. 2019a). Firstly, using the development of head loss across the first stage of a full-scale SWRO to monitor membrane performance is complicated by the fact that several types of fouling (particulate fouling and scaling) may occur simultaneously in SWRO membrane systems. Secondly, the use of intermittent non-oxidizing biocides to combat biofouling in full-scale SWRO membrane, makes establishing a real correlation between biological/organic fouling indicators in SWRO feed water and membrane performance very difficult. Thirdly, cleaning in place (CIP) may perform for various reasons than biofouling. Fourthly, many SWRO desalination plants with different pre-treatment processes in different parts of the world need to be monitored for long periods of time with different operating conditions. Regardless of these limitations, several attempts have been made to establish the relationship between biological/organic fouling indicators and membrane performance. Hijnen et al. (2009b) found that the pressure drop of membrane fouling simulator fed with fresh water depended on the AOC concentration present in the RO feed water, in which 1 μg-C/L (as acetate) added to the feed water of a membrane fouling simulator unit led to significant pressure drop in the RO membrane within 3 months. Weinrich et al. (2015b) reported obvious increase of differential pressure (0.28 -0.56 bar) within 4 months at 50 μg-C/L AOC concentration in the feed water of a pilot SWRO plant. Abushaban et al. (2019a) monitored BGP along the pre-treatment of three full-scale SWRO desalination plants and reported a preliminary correlation between BGP in SWRO feed water and the chemical cleaning frequency in SWRO.

Two main pre-treatment processes have been used to protect SWRO membranes from fouling; (i) conventional pre-treatment involving coagulation, flocculation, and particle separation and (ii) membrane filtration systems including microfiltration and ultrafiltration (Voutchkov 2010). The particle separation processes can consist of direct filtration with granular media, sedimentation and granular media filtration, and dissolved air flotation (DAF) and granular media filtration (Edzwald and Haarhoff 2011).

Media filtration has been widely used as a pre-treatment for SWRO systems either with or without inline coagulation. High removal of particulate, biological, and organic fouling potential has been reported by media filtration. Bonnelye et al. (2004) studied the removal of SDI in a pilot SWRO plant in the Gulf of Oman (open intake) and reported that a single stage of DMF combined with 1 mg-Fe^{3+}/L decreased SDI from 15 to less than 3.3 %/min. Abushaban et al. (2019b) measured microbial ATP along the pre-treatment of a full-scale SWRO desalination plant and reported more than 95 % removal in dual media filtration (DMF) combined with 1.3 mg-Fe^{3+}/L. Abushaban et al. (2019a) monitored BGP along the pre-treatment of three full-scale desalination plant, and found the highest removal (> 50 %) of BGP in DMF in combination with 0.8 – 3.6 mg- Fe^{3+}/L. Similarly, Weinrich et al. (2015b) reported low AOC concentration (1 – 150 µg-C/L) in the effluent of a media filter, which later increased in RO feed water due to chemical addition (Badruzzaman et al. 2019).

The DAF process has been coupled with granular media filtration processes in a number of SWRO desalination plants (Jacangelo et al. 2018). Kim et al. (2011) suggested not to use DAF alone as pre-treatment for SWRO due to limited particle removal, but rather DAF needs to be coupled with DMF to improve the pre-treatment performance. Kim et al. (2011) also found that the combination of DAF with DMF further reduced the particulate fouling potential, in which SDI_{15} and turbidity were 5.7 %/min and 0.25 NTU in the filtrate of DMF (without DAF) and decreased to 4.7 %/min and 0.17 NTU when DAF was coupled with DMF. However, insignificant organic matter removal was observed when DAF was coupled with DMF. Simon et al. (2013a) studied the removal of organics in a DAF-DMF pilot plant (coagulant dosage added into DAF system is not mentioned) located at El Prat de Llobregat (Barcelona, Spain) and reported low removal (12 % of dissolved organic carbon (DOC), 33 % of biopolymers, 0 % humic substances, 3 % of building blocks, and 10 % of low molecular weight acid (LMW-A)). Moreover, Abushaban et al. (2019a) also reported low reduction of BGP (15 %) in a DAF system (using 0.5 mg-Fe^{3+}/L) in a full-scale SWRO desalination plant located in the Middle East. Shutova et al. (2016) optimized the removal of organics in a seawater DAF system and reported optimum dosage of coagulant between 0.2 mg-Fe^{3+}/L (at pH 5.5) and 3.5 mg-Fe^{3+}/L (at pH 7.5).

Petry et al. (Petry et al. 2007) studied the effectiveness of DAF coupled with coagulation (coagulant dosage is not mentioned) prior to two-stages of DMF (El Coloso SWRO plant in Antogofasta, Chile) and reported low SDI_{15} values (< 3 %/min) in SWRO feed even when frequent algal bloom events occurred in the raw seawater. In another study, Faujour

115

et al. (2015) reported SDI_{15} values between 2 and 4 %/min in SWRO feed water at the Fujairah (II) SWRO desalination plant, in which DAF is coupled with 5 - 6.5 mg-Fe^{3+}/L coagulation/flocculation and gravity DMF. However, very little data is available on the removal of biological/organic fouling potential in SWRO pre-treatment, particularly, in full-scale SWRO desalination plants.

This research aims to investigate the relationship between the BGP of SWRO feed water and the pressure drop increase and permeability decline in SWRO system. For this purpose, biological/organic as well as particulate fouling indicators are used to monitor the pre-treatment of a full-scale SWRO desalination plant including DAF coupled with inline coagulation (1.0 - 1.6 mg- Fe^{3+}/L) and two stages of pressurized DMF. The SWRO plant was monitored for five months in terms of turbidity, total iron concentration, microbial ATP, particulate fouling potential (SDI and MFI), and organic indicators (total organic carbon (TOC), liquid chromatography analysis (LC-OCD)) and biological fouling potential (BGP and phosphate concentration). This work also presents information on the removal of biological/organic fouling potential through the pre-treatment of SWRO, in particularly in DAF-DMF seawater systems.

5.2 MATERIAL AND METHODS

5.2.1 Description of SWRO plant

The study was performed at a full-scale SWRO desalination plant fed via an open intake with seawater from the Gulf. Figure 5.1 shows the treatment scheme of the plant which consists of DAF combined with inline coagulation (1 - 5 mg-Fe^{3+}/L, depending on the SDI of the raw seawater), inline coagulation (0.3 – 1.5 mg-Fe^{3+}/L), two stage DMF, cartridge filtration (CF) with 5 μm pore size, and RO membranes. Phosphonate antiscalant dosed after the CF. The properties of the DMFs are presented in Table 5.1. DMFs are backwashed using SWRO brine (from the first pass).

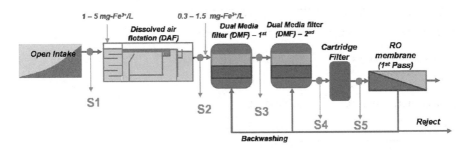

Figure 5.1: Schematic of the SWRO desalination plant with added coagulant dosage during the tested period (July – December).

5.2.2 Sample collection, measurement and transportation

Seawater samples were collected from the main header of the seawater intake (S1), after DAF (S2), after the first stage of dual media filtration (DMF1, S3), after second stage of dual media filtration (DMF2, S4), and after CF (S5). The properties of all collected seawater samples from intake and potable water are listed in Table 5.2. The following indicators were measured for 5 months; turbidity, total iron, microbial ATP, particulate fouling potential indicators (SDI-15 and MFI-0.45), biological fouling potential (BGP and orthophosphate concentration) and organic indicators (such as TOC and LC-OCD).

Table 5.1 Characteristics and operational properties of the two stage dual media filters.

	1st stage of DMF	2nd stage of DMF
No. and type of filters	24 horizontal pressure filters	16 horizontal pressure filters
Surface area	51 m²	51 m²
Filtration rate	12.5 m/h	19.5 m/h
Filtering media	0.55 mm sand and 1.50 anthracite	0.28 mm sand and 1.2 anthracite
Filtration cycle duration	~24 h	>40 h

Table 5.2: The water properties of influent and potable water

Parameter	Feed water	Potable water
pH	8.1-8.3	6.8-7.1
Turbidity	0.8-2.8 NTU	0.01-0.06 NTU
TDS	49–50 g/L	160–200 mg/L
Temperature	30-40 °C	30-40 °C
Boron		1.1 – 1.7 mg/L

5.2.3 Water quality characteristics

• Silt density index and modified fouling index

The standard methods in the American Society for Testing and Material (ASTM) to measure particulate fouling potential in RO system were used (namely; SDI and $MFI_{0.45}$). SDI is the rate of plugging of a membrane filter having 0.45 μm pores at a pressure of 210 kPa (30 psi) for a certain period of time. Typically, SDI of 15 min (SDI_{15}) is used. It should be known that the reported value should not exceed 75 % of the maximum value (5 %/min) (ASTM 2002). In case of high particulate fouling potential, shorter time needs to be used such as 10 min (SDI_{10}) or 5 min (SDI_5). If the reported value exceeds 75 % of SDI_5 (15), then $MFI_{0.45}$ should be used (ASTM 2002). For this study, SDI_5 was measured in the seawater intake and SDI_{15} was measured along the pre-treatment (after DMF1, DMF2 and CF). SDI and $MFI_{0.45}$ were measured using the portable SDI /MFI Analyzer (Convergence, Netherlands).

- ## **Microbial ATP**

The ATP filtration method was used to measure microbial ATP along the pre-treatment of the SWRO plant, which is described in Abushaban et al. (Abushaban et al. 2019b). Shortly, (i) seawater samples were filtered through sterile 0.1 μm PVDF membrane filters. (ii) The retained microorganisms on the membrane filter surface was rinsed with 2 mL of sterilized artificial seawater water. (iii) 5 mL of Water-Glo lysis reagent (Promega Corp., USA) was filtered through the filter to extract the microbial ATP from the retained cells. (iv) ATP of the filtrate was measured by mixing 100 μL aliquot with 100 μL of ATP Water-Glo detection reagent. The average emitted light measured by the Luminometer (GloMax®-20/20, Promega Corp.) was converted to microbial ATP concentration based on a calibration curve. Microbial ATP was measured on site. For each sample, six replications were measured.

- ## **Bacterial growth potential (BGP)**

Seawater samples were pasteurised (70 °C for 30 min) on-site to inactivate marine microorganisms and shipped to IHE Delft facilities (Delft, Netherlands) for analysis. All samples were collected in AOC-free 100 mL Duran® laboratory glass bottles with tight-fitting screw caps and transported in a cooler at 5°C within 36 h. BGP was measured following the described method by Abushaban et al. (2018). Shortly, the pasteurized sample was distributed in triplicate in 30 mL carbon-free vials and each vial was inoculated with 10,000 cells/mL (intact cell concentration measured by flow cytometry) of an indigenous microbial consortium. Samples were incubated at 30 °C and bacterial growth was monitored using microbial ATP measurement in seawater for 5 days Abushaban et al. (2019b).

- ## **Liquid chromatography - Organic Carbon Detection**

LC-OCD was used to measure the chromatography dissolved organic carbon (CDOC) and organic fractions including biopolymers, humic substances and low molecular weight (LMW) acids. The LC-OCD system separates dissolved organic carbon (DOC) compounds using a size exclusion chromatography column, followed by multi detection of organic carbon, UV-absorbance at 254 nm (UV_{254}) and Nitrogen (DOC-Labor, Germany). The measurement and analysis of the samples were performed according to the protocol described by Huber et al. (2011). Seawater samples were shipped in a cooler box (5 °C) to Doc-Labor Huber lab (Karlsruhe, Germany) for analysis.

- ## Total organic carbon (TOC)

TOC concentration in seawater was measured using a Shimadzu TOC-VCPN analyzer based on combustion catalytic oxidation/NDIR method.

- ## Orthphosphate

Orthophosphate analysis was performed using Skalar San^{++} analyzer (Skalar, Netherlands) at the facility of Rijkswaterstaat (Lelystad, Netherlands). Molybdate reagent and ascorbic acid were added to the seawater samples at a temperature of 37 °C. The added molybdate and the orthophosphate present in seawater samples form a phosphor-molybdate complex in the acidic environment after reduction with ascorbic acid and in the presence of antimone. This gave a blue colored complex, which was measured at 880 nm using a 50 mm cuvette and a spectrophotometer. The limit of detection of the orthophosphate analysis is 0.3 µg/L.

5.3   RESULT

5.3.1 Turbidity

Turbidity after the intake ranged between 0.5 and 2.9 NTU (Table 5.3). The highest turbidity (~ 2.9 NTU) in the seawater was measured in August, which was also confirmed by the SDI. The measured turbidity after DMF2 and CF were very low (< 0.1 NTU), indicating that most of the colloidal particles were removed through the two stages of media filtration. The removal of turbidity in DMF is also consistent with the reported values in the literature (Kim et al. 2011, Shrestha et al. 2014, Sabiri et al. 2017). Overall, more than 90 % of turbidity was removed along the pre-treatment of SWRO (from S1 to S5).

Table 5.3: Turbidity, SDI and MFI$_{0.45}$ along the pre-treatment of the SWRO plant over a period of 5 months (n= 20 samples).

		Seawater intake	After DMF1	After DMF2	SWRO feed	Overall removal
Turbidity	**Min.**	0.4	NA	<0.1	<0.1	0.3
(NTU)	**Max.**	2.9	NA	0.2	0.2	2.6
	Mean	1.5	NA	<0.1	<0.1	1.4 ± 0.9
SDI	**Min.**	9	3.5	2.8	2.6	6
(%/min)	**Max.**	>15	5.2	3.9	<4.0	> 11
	Mean	>15	4.4 ± 0.5	3.3 ± 0.4	3.2 ± 0.7	> 11
MFI	**Min.**	22	1.6	1.5	0.6	22
(s/L^2)	**Max.**	60	4.4	2.1	1.8	59
	Mean	41 ± 20	3.4 ± 1.2	1.7 ± 0.3	1.3 ± 0.5	39.7 ± 20

5.3.2 Particulate fouling potential indices

- ### SDI

High SDI values were measured (Table 5.3) in the seawater intake during the summer (July and August), which are above the maximum limit (SDI$_5$ = 15 %/min) defined by ASTM (ASTM 2002). The measured SDI$_{15}$ after DMF1 ranged between 3.5 and 5.2 %/min (with an average of 4.4 %/min) and further decrease after passing through DMF2 to 3.3 %/min. The measured SDI_{15} after DMF is close to the reported values by Bonnelye et al. (2004), who reported SDI_{15} below 3.3 %/min after DMF. Some literature reported even higher SDI_{15} (> 6.6 %/min) after DMF (Sabiri et al. 2017). As expected, negligible improvement in SDI_{15} was observed through the CF. Overall, the measured SDI_{15} after CF was below the recommended SDI_{15} values (< 4 %/min) by the membrane manufacturers, indicating low particulate fouling potential in the SWRO feed water.

- **$MFI_{-0.45}$**

High $MFI_{-0.45}$ variations were observed in the seawater intake, these values ranged between 22 to 60 s/L^2 (Table 5.3). The measured $MFI_{-0.45}$ values are lower than the reported values by Salinas Rodriguez et al. (2019) in the raw seawater of North Sea (20 - 250 s/L^2), suggesting lower particulate fouling potential in the monitored SWRO plant. Similar to SDI, significant removal of $MFI_{-0.45}$ was observed in DMF1 and DMF2, in which $MFI_{-0.45}$ decreased from 41 s/L^2 in the seawater intake to 3.4 s/L^2 after DMF1 and to 1.7 s/L^2 after DMF2. Shrestha et al. (2014) reported slightly higher $MFI_{-0.45}$ values after lab-scale sand filters (1.9–5.9) even though the $MFI_{-0.45}$ of influent was much lower (4 – 10 s/L^2). Slight improvement of $MFI_{-0.45}$ was found after the cartridge filter, in which the average measured $MFI_{-0.45}$ in the SWRO feed water was 1.3 s/L^2. Overall, 97 % removal of $MFI_{-0.45}$ was achieved in the SWRO pre-treatment.

5.3.3 Biomass quantification

Microbial ATP concentration in the seawater intake varied from 75 to 335 ng-ATP/L (Fig. 5.2). High microbial ATP concentrations (>> 100 ng-ATP/L) were observed in July and August, which could be attributed to microbial growth as a result of high water temperature (32 - 40 °C) in July and August. Whereas, from September to December, microbial ATP concentrations fluctuated around 100 ng-ATP/L. This was also observed in the North Sea water by Abushaban et al. (2018) who reported high seasonal variations in microbial ATP concentrations ranging between 25 and 1,000 ng-ATP/L.

Monitoring microbial ATP through the pre-treatment showed that, on average, 27 % of microbial ATP was removed through the DAF system, in which microbial ATP concentrations after DAF ranged between 50 and 170 ng-ATP/L. Significant removal of microbial ATP (60 %) was found in DMF1 in combination with inline coagulation (0.3 –

1.5 mg-Fe^{3+}/L). This is close to the reported removal (65 – 85 %) in a pilot seawater media filter (without coagulation) fed with seawater from the North Sea (Abushaban et al. 2019b). Further removal of microbial ATP was seen in the DMF2 (45 %) and the CF (16 %). Microbial ATP concentration in the SWRO feed water ranged between 10 and 35 ng-ATP/L. In total, more than 86 % of microbial ATP was removed through the SWRO pre-treatment. Abushaban et al. (2019b) reported higher removal of microbial ATP (95 %) in a full-scale SWRO plant with two stages of DMFs (with similar properties of the DMFs of the monitored plant) coupled with inline coagulation (1.3 mg- Fe^{3+}/L). The higher removal of microbial ATP is attributed to the higher coagulant dosage prior to DMF in this study.

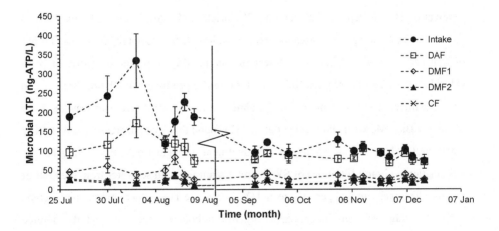

Figure 5.2: Microbial ATP concentrations along the SWRO pre-treatment over a period of 5 months (n=18 samples). Intake (the seawater intake), DAF (After dissolved air flotation), DMF1 (after the first stage of dual media filtration), DMF2 (after the second stage of dual media filtration) and CF (after cartridge filtration).

5.3.4 Organic indices

- ## Total organic carbon

High TOC concentration was measured in the seawater intake ranging between 1.9 and 5 mg/L (Table 5.4) with an average of 2.9 mg/L. After DAF and DMF1, the TOC concentration declined to 2.3 mg/L (15 %) and 2.0 mg/L (13 %), respectively. The removal is close to that reported in literature. Shutova et al. (2016) reported 16 % removal of DOC in a lab scale DAF system fed with Gold Coast seawater and with 1 mg-Fe^{3+}/L. Jeong et al. (2016) observed 0.1 mg/L (12%) removal of DOC in the DMF of Perth SWRO desalination plant. Slight TOC removal was found through DMF2 (5 %) and after CF (6 %)

The overall removal of TOC along the pre-treatment is 33 %. However, even lower removal of TOC is reported in literature. Weinrich et al. (2011) reported only 3 - 6 % removal of TOC along the pre-treatment (coagulation (dosage is not reported), sand filter, diatomaceous filter and cartridge filter) of Tampa Bay seawater desalination plant (Florida, USA) and no removal of TOC through the pre-treatment (Ultrafiltration and cartridge filter) of a pilot plant in Moss Landing (California, USA) fed with seawater from Monterey Bay. Morover, Poussade et al. (2017) found that TOC decreased from 1.14 to 0.89 mg/L (13.5 %) through the pre-treatment (coagulation with 1 mg-Fe^{3+}/L, flocculation and sand filtration) of SWRO pilot plant fed with seawater from the Gulf of Oman. The lower removal percentage of TOC is because TOC concentration may include high percentage of non-biodegradable organic carbon and the fact that the applied coagulant dosage was low.

- ## LC-OCD fractions

High concentration of organic fractions including hydrophilic dissolved organic carbon (CDOC), biopolymers, humic substances and low molecular weight acid (LMW acid) was observed in summer (July and August) compared to the measured fractions in the Autumn (September to December). In total, 384 μg-C/L of CDOC was removed (21 %) through the SWRO pre-treatment (Table 5.4). The highest removal of CDOC was measured in the DAF and DMF1 (7 % and 9 %, respectively). The CDOC removal in DAF (135 μg-C/L) was mainly due to the removal of humic substances (78 μg-C/L) and biopolymers (67 µg-C/L). Similar findings were reported at bench-scale DAF system byShutova et al. (2016). The high removal of humic substances in the DAF was also confirmed by the monitored FEEM (See annex S5.A).

Even though the highest CDOC removal was found in DMF1 (143 µg-C/L), removal of biopolymers, humic substances and LMW acids observed in DMF1 (21, 8, and 9 µg-C/L, respectively) was low. Slightly higher removal of biopolymers, humic substances and LMW acids were observed in DMF2 (38, 16 and 5 µg-C/L, respectively) than DMF1, probably due to smaller media size in DMF2 (Table 5.1). The low removal of humic substances in DMF2 was expected as humic substances are mainly removed by coagulation. Shrestha et al. (2014) reported only 2 % removal of humic substances in sand and anthracite biofilters. CF showed no removal of organic carbon, as expected. Overall, low removal of organic fractions was seen through the pre-treatment of the SWRO desalination plant with the best removal in the DAF system

5.3.5 Biofouling indicators

- ### Orthophosphate

The orthophosphate concentration measured in the seawater intake ranged between 2 and 11 µg-P/L (Table 5.5). Munshi et al. (2005) measured orthophosphate concentration in the raw seawater (Arabian Gulf) and the permeate of nano-filtration of Al-Jubail SWRO desalination plant and reported 4.7 and 1.1 µg-P/L, respectively. Significant removal (68 %) of orthophosphate was observed through DAF and further removal (33 %) was found through the DMF1. The high removal of phosphate in the DAF and DMF1 could be attributed to the precipitation of iron phosphate as coagulant 1 - 5 mg-Fe^{3+}/L was added prior to DAF and DMF1 (Nir et al. 2009). It is worth mentioning that no data is available in the literature on the removal of orthophosphate in the pre-treatment processes of SWRO membrane systems. Similar to BGP and TOC, orthophosphate concentration increased after CF from 1.1 to 1.5 µg-P/L, which may be attributed to the addition of phosphonate antiscalant and/or to the presence of nutrients in the make-up water.

Table 5.4: Removal of various fractions of organic carbon (LC-OCD) along the pre-treatment of SWRO desalination plant over 5 months period (n = 5 samples).

Parameter		Seawater intake	After DAF	After DMF1	After DMF2	After CF	Overall removal
Coagulation (mg-Fe^{3+}/L)		-	1 – 5	0.3 – 1.5	-	-	
TOC (mg/L)	Mean	2.7 ± 0.8	2.3 ± 0.3	2.0 ± 0.2	1.9 ± 0.2	1.8 ± 0.1	0.9 ± 0.6
	(%removal)		(15 %)	(13 %)	(5 %)	(6 %)	(33 %)
CDOC (µg-C/L)	Min.	1,543	1,409	1,400	1,317	1,236	307
	Max.	2,026	1,911	1,589	1,711	1,679	573
	Mean	1,808 ± 244	1,673 ± 268	1,530 ± 90	1,468 ± 174	1,424 ± 190	384 ± 127
	(%removal)		(7 %)	(9 %)	(4 %)	(3 %)	(21 %)
Biopolymers (µg-C/L)	Min.	216	165	160	120	126	89
	Max.	339	236	196	152	149	192
	Mean	265 ± 57	198 ± 35	177 ± 19	140 ± 15	141 ± 10	124 ± 51
	(%removal)		(25 %)	(11 %)	(21 %)	(0 %)	(47 %)
Humic subs. (µg-C/L)	Min.	577	529	540	511	481	58
	Max.	881	796	764	755	755	143
	Mean	737 ± 165	660 ± 147	651 ± 125	635 ± 132	623 ± 143	114 ± 38
	(%removal)		(10 %)	(1 %)	(2 %)	(2 %)	(15 %)
LMW- acid (µg-C/L)	Min.	115	121	115	106	102	4
	Max.	203	192	183	181	175	35
	Mean	157 ± 47	157 ± 37	149 ± 36	144 ± 39	139 ± 38	18 ± 13
	(%removal)		(0 %)	(5 %)	(3 %)	(3 %)	(11 %)

Table 5.5: Monitored orthophosphate and BGP along the pre-treatment of the SWRO desalination plant over a 5 months period (n = 12 samples).

Parameter		Seawater intake	After DAF	After DMF1	After DMF2	After CF
Coagulant dosage (mg-Fe^{3+}/L)		-	$1-5$	$0.3-1.5$	-	-
Phosphate (μg-C/L)	Min.	1.8	1.0	0.6	0.7	1.1
	Max.	11	2.6	1.5	1.5	2.6
	Mean	5.3 ± 3.7	1.7 ± 0.6	1.1 ± 0.4	1.1 ± 0.2	1.5 ± 0.6
	(% removal)		(68 %)	(35 %)	(0 %)	(-36 %)
BGP (μg-C/L)	Min.	200	112	72	65	55
	Max.	1,030	403	203	148	388
	Mean	373 ± 268	180 ± 61	106 ± 32	92 ± 25	146 ± 106
	(% removal)		(52 %)	(40 %)	(14%)	(-37 %)

- **Bacterial growth potential**

High BGP variations were observed in the seawater intake, in which BGP ranged between 200 and 2,500 μg-C/L as glucose (Fig. 5.3). Extremely high BGPs were observed from the end of August to October in the seawater intake and along the pre-treatment due to algal blooms. Algal blooms in the Arabian Sea in September and October is widely reported (Piontkovski et al. 2011, Sarma et al. 2012). It is believed that higher BGP in the summer might be attributed to carbon release from the algal cells present in seawater.

The highest BGP removal was found through DAF (52 %) and DMF1 (40 %). This result is in agreement with the findings of Kim et al. (2011) who reported similar removal of organic fractions in terms of chemical oxygen demand (35 %), UV_{254} (23 %) and chlorophyll A (45 %) in both DAF and DMF when combined with inline coagulation (1.3 mg-Fe^{3+}/L) The high removal of BGP in DAF could be attributed to the coagulant dosage within DAF ($1-1.6$ mg-Fe^{3+}/L), while the achieved removal of BGP in DMF1 may be due to the applied inline coagulant dosage (0.35 mg-Fe^{3+}/L) prior to DMF1 and/or due to biodegradation in DMF1. Abushaban et al. reported slightly higher BGP removal where 44 % in a pressurized pilot media filter without coagulation dosage and 55 % in a gravity

DMF combined with inline coagulation (3.6 mg-Fe^{3+}/L) were obtained (Abushaban et al. 2018, Abushaban et al. 2019b)

Slight removal (14 %) of BGP was also noted through DMF2, which may be due to the shorter contact time compared with DMF1 and/or the absent of coagulant dosage. One may expect higher organic biodegradation in DMF2 because the filtration cycle of DMF2 is longer (> 40 h) compared to DMF1 (~ 24 h). The long filtration time may allow the development of a substantial biofilm on the filter media. However, the use of SWRO brine to backwash DMF may have hindered the initial formation of biofilm due to the osmotic shock expressed by the bacteria. This has been verified by monitoring microbial ATP in the filtrate of the DMFs (See annex S5.B). These results could suggest an impact of using SWRO brine to backwash media filters on biofilm development.

Higher BGP was observed after the CF which could be attributed to the addition of antiscalant (Vrouwenvelder et al. 2000) or the make-up water used for diluting antiscalant (See annex S5.C). The higher organic concentration after antiscalant addition has been observed in several SWRO and RO plants (Schneider et al. 2005, Jeong et al. 2016). On average, BGP was reduced from 373 µg-C/L (as glucose equivalent) in the seawater intake to 146 µg-C/L (as glucose equivalent) in the SWRO feed water. The removal of BGP (62 %) along the SWRO pre-treatment is comparable to the reported BGP removal of 50 - 72 % by Abushaban et al. (2019a) in three full-scale SWRO desalination plants with different pre-treatment processes.

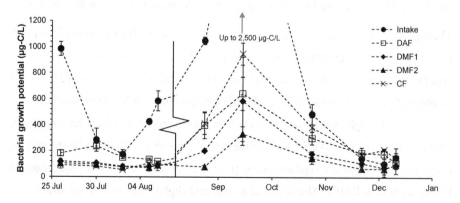

Figure 5.3: BGP along the pre-treatment of the SWRO desalination plant.

5.4 DISCUSSION

Several parameters have been monitored along the SWRO pre-treatment and in the SWRO feed water including particulate, biological and organic fouling indicators over a 5 months period. It is assumed that scaling did not occur as antiscalant is dosed prior to the SWRO membranes and thus should eliminate the occurrence of any scale in the first pass of the SWRO plant.

- ### Turbidity

Significant removal of turbidity and iron was observed through the pre-treatment. The measured turbidity (< 0.1 NTU) and iron concentration (< 0.03 mg-Fe^{3+}/L) in the SWRO feed were below the recommended values (< 0.1 NTU and 0.05 mg-Fe^{3+}/L, respectively) according to the membrane manufacturer.

- ### Particulate fouling

Results showed that particulate fouling was well controlled through the pre-treatment. This can be justified by several observations. Firstly, high removal (> 80 %) of particulate fouling indices (SDI_{15} and $MFI_{0.45}$) was observed through the SWRO pre-treatment, in which the highest removal was achieved in DMF1 combined with 0.3 -1.5 mg-Fe^{3+}/L as an inline coagulation. The high removal in the DMF is in agreement with what was reported earlier by others (Peleka and Matis 2008, Salinas Rodriguez et al. 2019). Secondly, the measured SDI_{15} (3.2 ± 0.7) in the SWRO feed was below the manufacturers recommended values (< 4 %/min). Thirdly, applying the particulate fouling prediction model (based on MFI) presented by Salinas Rodriguez et al. (2019), the SWRO system can be operated for more than two years before observing a one bar increase in the net driving pressure of SWRO membrane system (See annex S5.D), for MFI < 2 s/L^2 in the SWRO feed water. Therefore, according to the measured turbidity (<0.1 NTU), SDI_{15} (3.2 %/min) and $MFI_{0.45}$ (1.3 s/L^2) in the SWRO feed, particulate fouling unlikely to be the main reason for fouling.

- ## Biomass quantification

Having high microbial concentration in the SWRO feed water does not directly cause biofouling. It may cause particulate fouling and/or accelerate bacterial growth in SWRO membrane system and thus indirectly may increase the rate of biofouling. Significant removal of microbial ATP (85%) was also observed through the SWRO pre-treatment (Fig. 5.2), in which microbial ATP concentration decreased, on average, from 130 ng-ATP/L in the raw seawater intake to 18 ng-ATP/L in the SWRO feed water. The microbial ATP concentration in the SWRO feed water is equivalent to 20,000 intact cells per mL (using the reported correlation between microbial ATP and intact cell concentration of North Sea water (Abushaban et al. 2019b)).

- ## Biological and organic fouling in the pre-treatment

Compared to the removal of particulate fouling potential and microbial ATP, lower removal percentages of biological/organic fouling potential was seen along the SWRO pre-treatment train. However, DAF combined with 1 - 5 mg-Fe^{3+}/L coagulant dosage showed reasonable removal of biological/organic fouling potential, in which 3.6 μg-P/L of orthophosphate (68 %), 197 μg-C/L of BGP (52 %), 77 μg-C/L of biopolymers (25 %), 135 μg-C/L of CDOC (7 %), and 77 μg-C/L of humic substances (10 %) were removed. Shutova et al. (2016) studied the removal of organic matter in DAF system, used as pre-treatment for SWRO membrane. The magnitude of removed organic fractions (Biopolymers: 60 - 65 μg-C/L, CDOC: 140 - 240 μg-C/L and humic substances: 100 - 180 μg-C/L) are in the same range to the observed removal in this study.

Good removal of biological/organic fouling potential was measured in DMF1 combined with 1 – 5 mg-Fe^{3+}/L of coagulation comparing to the reported removal in the literature. The observed removal of BGP (74 μg-C/L, 40 %), CDOC (143 μg-C/L, 9 %), biopolymers (21 μg-C/L ,10 %) and humic substances (9 μg-C/L, 1.3 %) in DMF1 were higher than reported Jeong et al. (2016) in the DMF of Perth SWRO desalination plant, in which they reported 13 % of AOC (5 μg-C/L), 6.6 % of CDOC (100 μg-C/L), 11 % of biopolymers (10 μg-C/L), and 0 % of humic substances. Moreover, the overall removal of organic fractions in the DAF and the DMF1 of the studied plant is higher than the reported organic removal by Simon et al. (2013a) after DAF (coagulant dosage not mentioned) and DMF of a pilot plant located at El Prat de Llobregat (Barcelona, Spain), in which 161 µg-C/L of CDOC (12 %), 35 µg-C/L of biopolymers (13 %), 0 µg-C/L of humic substances (0 %) and 6 µg-C/L of LMW-acid (10 %) were removed.

These results reveal that the achieved removal of biological/organic fouling potential in the monitored SWRO plant is comparable to SWRO plants at different locations, and even higher than some SWRO plants. However, even better removal of biological/organic fouling potential could be achieved by adjusting several design and operational parameters. For instance, extending the contact time of the DMF is expected to enhance biodegradation of organics. Moreover, the use of SWRO brine to backwash the media filters could burst the microorganisms/biofilm in the media filtration and thus affect the biodegradation rate in DMF, because the high osmotic pressure that the biofilm is exposed to (during backwashing).

- ### Biological/organic fouling in the SWRO feed

Although reasonable concentration of organic and biological fouling potential was removed through the pre-treatment, still considerable concentration remains in the SWRO feed water (Table 5.4 & 5.5). As no standard threshold value for organic and biological fouling potential is available, the measured concentration in the SWRO feed water is firstly compared with those reported in the literature. According to the literature, the fouling in the SWRO system is most likely due to biofouling for the following reasons; (i) Jeong et al. (2016) observed biofouling in the SWRO system at the Perth desalination plant where lower organic fractions (1.3 mg/L of CDOC, 50 µg-C/L of biopolymers, 140 µg-C/L of humic substances) in the SWRO feed water were found which are lower than those measured as organic fractions (1.4 mg/L of CDOC, 141 µg-C/L of biopolymers and 623 µg-C/L of humic substances). (ii) Weinrich et al. (2015b) reported a preliminary AOC threshold concentration of 50 µg-C/L based on pilot tests, while 146 µg-C/L of BGP was measured in the SWRO feed (assuming AOC and BGP are similar). Thus, it was suggested that biofouling in the SWRO membrane occurred due to high potential of biological/organic fouling in the SWRO feed water.

- ### Investigating the relation between membrane performance and BGP in SWRO feed water

The relationship between BGP in the SWRO feed water and the normalized pressure drop/permeability in the SWRO membrane system was studied (Fig. 5.4), and found that higher BGP measured from July to September, corresponded to a higher normalized pressure drop. The measured BGP in the SWRO feed water in July were all around 100 µg-C/L, during which time the normalized pressure drop further inclined and the normalized permeability also further declined, suggesting that 100 µg-C/L of BGP may still be sufficient to cause biofouling in SWRO membrane systems. This result suggests that BGP could be used to monitor biological fouling in SWRO system. However, more data need to be generated at different SWRO plants at different locations in order to validate the use of BGP as a biological fouling indicator.

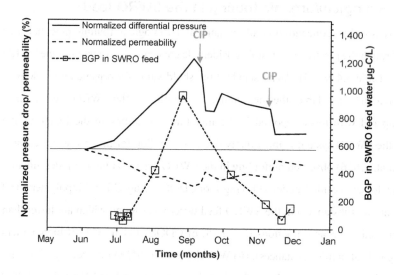

Figure 5.4: Correlation between BGP in the SWRO feed water and the normalized pressure drop and normalised permeability in the SWRO membrane system (n = 11).

5.5 CONCLUSIONS

- High seawater quality variations were observed in the seawater intake in terms of silt density index (SDI), modified fouling index (MFI), microbial ATP, bacterial growth potential (BGP), orthophosphate and total organic carbon.

- Particulate fouling was well controlled by the SWRO pre-treatment, in which the measured SDI-$_{15}$, MFI-$_{0.45}$ and turbidity in the SWRO feed water were all below the recommended values. The highest removal (70 – 90 %) of SDI-$_{15}$, MFI-$_{0.45}$ and turbidity was achieved in the first stage of Dual Media Filtration when combined with inline coagulation using 0.3 – 1.5 mg-Fe^{3+}/L.

- Despite removal of biological/organic fouling potential (> 75 %) along the SWRO pre-treatment, particularly in the dissolved air flotation and the first stage of Dual Media Filtration, BGP and orthophosphate concentrations increased by 35 % in the SWRO feed due to chemical addition, and/or due nutrients present in the water storage tanks or make-up water.

- Investigating the relation between normalized pressure drop in the SWRO system and Bacterial Growth Potential in the SWRO feed water showed that the growth potential measured in the SWRO feed water from 100 to– 950 µg-C/L led to an obvious increase in the normalized pressure drop within 3 months. This result may suggest the applicability of using Bacterial Growth Potential of SWRO feed water as a biological fouling indicator in SWRO systems. However, in order to ensure the validity of this conclusion, more SWRO plants need to be monitored at different locations for longer periods of time.

5.6 SUPPLEMENTARY MATERIALS:

- ## Annex A: Fluorescence excitation-emission matrix (FEEM) spectroscopy

The fluorescence emitting organic substances in seawater samples were measured using a FluoroMax-3 spectrophotometer (Horiba Jobin Yvon, Inc., USA) with a 150 W ozone-free xenon arc lamp to enable excitation. The seawater samples were scanned over the excitation wavelength range from 240 to 450 nm, and an emission wavelength range of 290 to 500 nm to produce a three-dimensional matrix. Before FEEM analysis, the dissolved organic carbon (DOC) of seawater samples was measured using Shimadzu TOC, and diluted with MilliQ water to get a DOC concentration of approximately 1 mg/L. Excitation and emission matrices were analyzed using MatLab R 2011a, and the results were interpreted as described by Leenheer and Croué (2003).

High peaks were observed above 380 nm of excitation indicating the presence of high fraction of humic- like (marine) and humic-like components (Fig S5.1). No clear peak of protein-like was seen (excitation range from 310 - 360 nm). Clear removal of humic-like compounds was observed through the pre-treatment, in particularly, through DAF and DMF1, in which the intensity decreased to half (from 2×10^{-3} to 1×10^{-3}). This is in agreement to the findings of Shutova et al. (2016) who reported $46 - 49$ % removal of humic-like substances in bench-scale seawater DAF system. Slight removal of humic-like component was also seen in DMF2. Similar trend was also seen for the fulvic acid-like substances but with higher intensity (more than 3.5×10^{-3}) indicating higher contribution of fulvic acid-like compounds than humic acid-like.

- ## Annex B: Hourly monitoring of microbial ATP over the day

Microbial ATP concentration was monitored over two days (July 27th and 28th) along the pre-treatment and significantly variation was found in the seawater intake (between 150 and 450 ng-ATP/L) (Fig. S5.2). Microbial ATP concentration after DAF was stable at 130 ng-ATP/L for the monitored 2 days showing the added value of DAF in stabilizing the seawater. Microbial ATP concentration ranged between 50 and 80 ng-ATP/L and between 5 and 25 ng-ATP/L before the backwashing of DMF1 and DMF2, respectively.

Immediately after backwashing, low microbial ATP concentrations was measured after DMF1 and DMF2 due to the use of high saline seawater (SWRO brine) to backwash the DMFs. High saline seawater could burst/kill the microorganisms and biofilm present in the media filters. Dramatic increase of microbial ATP concentration was observed during maturation period, which is attributed to the replacement of SWRO brine with the real seawater (less saline). The microbial ATP concentration measured after maturation was higher than the microbial ATP concentration measured before backwashing of DMF1 and DMF2 for (1 – 4 h) which may indicate that long maturation time after backwashing is needed.

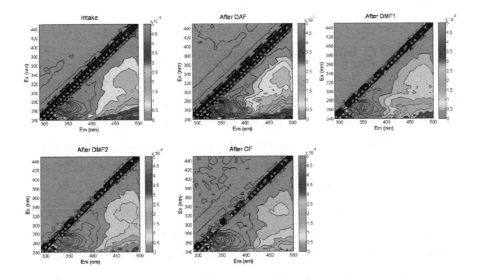

Figure S5.1: FEEM fluorescence features along the SWRO pre-treatment for seawater samples collected in August 2018

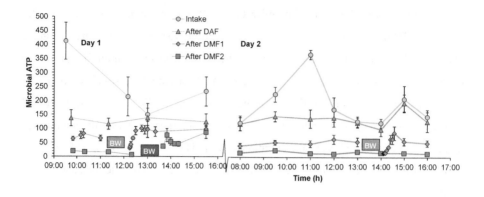

Figure S5.2: Hourly monitoring of microbial ATP concentrations for the seawater intake, after DAF, DMF1 and DMF2. BW = Backwashing time.

• Annex C: Effect of antiscalant addition

The BGP, TOC and phosphate concentration increased after the cartridge filter, in which the highest increase was 60 % in the BGP (from 92 to 146 µg-C/L). This could probably be attributed to addition of antiscalant after the cartridge filter. Therefore, the effect of antiscalant (similar to the applied dosage in the plant) was investigated by comparing BGP of a seawater sample with and without addition of the antiscalant. It was found that the BGP increased by 28 % comparing to the sample without antiscalant as shown in Table S5.1. This increment percentage (28 %) is lower than the measured increment after cartridge filtration (60 %) suggesting that antiscalant partially contributed in the higher biological/organic fouling indicators after cartridge filter. The rest 32 % might be due to contaminated water used for diluting the antiscalant in the full-scale SWRO desalination plant.

Table S5.1: the effect of antiscalant addition on BGP measurement

	BGP (µg/L)
BGP without Antiscalant	150
BGP with Antiscalant	192

- ## Annex D: Operating time of RO membrane system based on MFI in RO feed water

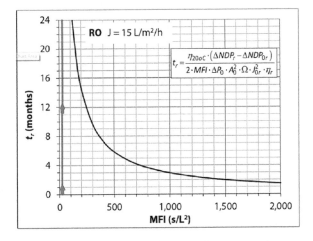

Figure S5.3: The expected time of particulate fouling in RO membrane system calculated based on MFI value in the RO feed water (Salinas Rodriguez et al. 2019).

6

APPLICATION OF ATP AND BGP METHODS TO MONITOR MEDIA FILTRATION AND DISSOLVED AIR FLOTATION PRE-TREATMENT SYSTEMS

Media filtration is the most common used pre-treatment process in full-scale SWRO desalination plants. The removal of particulate fouling in seawater media filtration has been extensively addressed. However, the removal of organic/biological fouling potential in seawater media filtration is less extensively covered in the literature, mainly due to a lack of standard methods for these types of fouling. Recently, several biological/organic fouling indicators have been developed such as assimilable organic carbon and bacterial growth potential (BGP). This work aims (i) to investigate the removal of particulate, organic and biological fouling potential during two stage dual media filtration (DMF) pre-treatment in a full-scale SWRO desalination plant and (ii) to compare the removal of fouling potential in two stage DMF (DMF pre-treatment) with the removal in two stage DMF installed after dissolved air floatation (DAF-DMF pre-treatment). For this purpose, the silt density index (SDI), modified fouling index (MFI), BGP, organic fractions and microbial ATP were monitored through the pre-treatment processes of two full-scale SWRO desalination plants.

Particulate fouling potential was well controlled through the two stages of DMF with significant removal of SDI_{15} (80 %), $MFI_{-0.45}$ (94 %), and microbial ATP (> 95 %). However, lower removal of biological and organic fouling potential (24 – 41 %) was observed compared to the observed removal of particulate fouling, which was attributed to low biological activity in the DMFs due to frequent chlorination (weekly), which probably destroyed the biofilm layer on the filter media.

Comparing overall removal of DAF-DMF pre-treatment to DMF pre-treatment showed that DAF significantly improved the removal of biological/organic fouling potential, in which the removal of BGP and biopolymer concentration increased by 40 % and 15 %, respectively. However, the removal of biological/organic fouling potential in the two stage DMFs in both plants were in the same range in terms of BGP (90 - 95 µg-C/L), CDOC (320 - 460 µg-C/L), and biopolymers (100 - 160 µg-C/L). The low removal of biological/organic fouling potential in both stages of DMF (in DAF-DMF) was attributed to the periodic use of SWRO brine for backwashing, which probably resulted in osmotic shock of the biofilm on the filter media, and reduced removal of biodegradable organic matter.

Overall, relatively low removal of biological/organic fouling potential was observed in DMFs in two full-scale SWRO plants due to operational practices such as chlorination of the intake and brine backwashing. Monitoring ATP and BGP along the pre-treatment processes, particularly in DMF, would be beneficial to enhance biological degradation and lower the BGP of SWRO feed water.

Keywords: Desalination, fouling potential, seawater reverse osmosis, pre-treatment, seawater monitoring.

This chapter is based on **Almotasembellah Abushaban**, Sergio G. Salinas-Rodriguez, Delia Pastorelli, Jan C. Schippers, and Maria D. Kennedy. Application of ATP and BGP methods to monitor media filtration and dissolved air flotation pre-treatment systems. In preparation to be submitted to Desalination.

6.1 INTRODUCTION

The key operational challenge that seawater reverse osmosis (SWRO) systems face during operation is membrane fouling (Matin et al. 2011, Goh et al. 2018). Membrane fouling reduces membrane permeability; permeate quality and leads to higher operating pressures. Membrane fouling can be due to suspended and colloidal particles, organic matter, dissolved nutrients and sparingly soluble salts. One or more types of fouling can occur depending on feed water quality, operating conditions and type of membrane (She et al. 2016). To mitigate fouling in SWRO systems, pre-treatment is commonly applied (Henthorne and Boysen 2015). Pre-treatment improves the quality of SWRO feed and increases the efficiency and life expectancy of the membrane elements by minimizing fouling, scaling and degradation of the membrane (Dietz and Kulinkina 2009).

Pre-treatment includes physical and chemical water treatment processes. Typically, two types of pre-treatment systems are used to protect SWRO membranes from fouling; (i) conventional pre-treatment including coagulation followed by granular media filtration and (ii) advanced treatment including microfiltration or ultrafiltration (Voutchkov 2010). Conventional pre-treatment is still the most common in full-scale SWRO desalination plants (Valavala et al. 2011).

The removal efficiency of the pre-treatment is currently monitored by turbidity and particulate fouling indices such as the silt density index (SDI) and modified fouling index (MFI). The recommended SDI and turbidity (by membrane manufacturers) to avoid particulate fouling in RO membrane systems are 4 %/min and 0.5 NTU, respectively. However, no standard method and threshold values exist to monitor and control biological/organic fouling potential in SWRO feed water. Consequently, several methods to monitor organic/biological fouling potential have been developed (Weinrich et al. 2013, Jeong et al. 2016).

It has been reported that conventional pre-treatment can remove low percentage of organic matter. Weinrich et al. (2011) reported only 3 - 6 % removal of TOC along the pre-treatment (coagulation (dosage is not reported), sand filter, diatomaceous filter and cartridge filter) in Tampa Bay seawater desalination plant (Florida, USA). Moreover, Poussade et al. (2017) found that TOC decreased from 1.14 to 0.89 mg/L (13.5 %) through the pre-treatment (coagulation with 1 mg-Fe^{3+}/L, flocculation and sand filtration) of SWRO pilot plant fed with seawater from the Gulf of Oman. Furthermore, Jeong et al. (2016) observed 0.1 mg/L (12%) removal of DOC in the DMF of Perth SWRO desalination plant. However, some other studies reported high removal of organic matter such as Shrestha et al. (2014) who observed 45 % removal of DOC and biopolymers and 62 % removal of low molecular weight acid in a seawater sand filter. Kim et al. (2011) observed 50 % removal of organic matter as UV_{254} absorbance and 56 % removal of chlorophyll a in a seawater DMF pilot plant in South Korea.

To minimize biofouling and organic fouling, the use of biological filtration as a pre-treatment of SWRO systems was suggested by Naidu et al. (2013) and Fonseca et al. (2001). In biological filtration, organic matter can be biodegraded or adsorbed on media. Shrestha et al. (2014) compared the removal efficiency of three biofilter columns, with granular-activated carbon (GAC), anthracite and sand as a biofilter media and reported high removal efficiencies in the three biofilters in terms of turbidity (around 50 %), modified fouling index (50 – 80 %) and hydrophobic organic compounds (94 %). In addition, the removal of DOC was 41 – 88 % with GAC media, 7 –76 % sand media and 3 – 71 % with anthracite media.

In this chapter, the removal of fouling potential in a full-scale SWRO desalination plant with two stage DMF coupled with inline coagulation pre-treatment was monitored in terms of particulate, biological and organic fouling potential. Additionally, the removal of fouling potential in two stage DMF pre-treatment was compared with the removal in two stage DMF preceded by dissolved air flotation (DAF-DMF).

6.2 MATERIALS AND METHODS

6.2.1 Description of SWRO plant

The study was performed at a full- scale seawater desalination plant. Figure 6.1 shows the treatment scheme of the SWRO plant which consists of inline coagulation, two stages of DMF, cartridge filtration, RO membrane (2 pass) and remineralization.

Seawater is collected through an inlet pipe (open intake) located 1.3 kilometres from the shoreline. Sodium hypochlorite (1 mg-Cl_2/L) was dosed weekly for 4 hours at the intake basin. During chlorination, sodium bisulphite is dosed prior to the cartridge filtration to quench the residual chlorine and to prevent oxidation of the RO membranes.

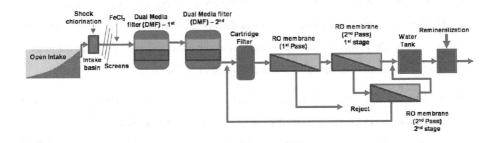

Figure 6.1: Schematic of the SWRO desalination plant (plant A).

The main pre-treatment of the SWRO system is two stage DMFs operated in series. Table 6.1 shows the properties of each stage. Prior to the DMF, inline coagulation is applied using 0.7 – 1.7 mg- Fe^{3+}/L. Backwash of the filter was performed using the filtered seawater from the second stage of DMF. After DMF, seawater is filtered through cartridge filtration (5 µm) to remove any sand and particles coming from the DMF. In addition, 1.9 mg/L of antiscalant is dosed before the cartridge filtration.

RO membranes are arranged in a two pass array. The first RO pass consists of 14 trains operating at 40 % RO recovery. The second pass consist of seven trains (two stages) operating at 90 % RO recovery. The total recovery of the first and second pass is 40 - 43 %.

Table 6.1: Characteristics and operational properties of the two stages media filtration.

	1st stage of DMF	2nd stage DMF
No. and type of filter	23 horizontal pressure filters	16 horizontal pressure filters
Filtration rate	12.5 m/h	19.5 m/h
Media size	0.95 mm sand and 1.5 mm anthracite	0.28 mm sand and 0.95 mm anthracite
Filtration period	~24 h	120 h

6.2.2 Sample collection, measurement and transportation

Seawater samples were collected over four days (from the main header pipe of the seawater intake), after the first stage of dual media filtration, after second stage of dual media filtration, and after cartridge filtration. Chlorination was applied on the fourth day whereby 1 mg-Cl_2/L was dosed. The properties of the collected seawater from the intake, the pre-treatment train and potable water are listed in Table 6.2.

Microbial adenosine triphosphate (ATP), bacterial growth potential (BGP), biopolymer concentration, hydrophilic dissolved organic carbon (CDOC), SDI and $MFI_{-0.45}$ were measured. Samples were collected in carbon-free bottles for microbial ATP, BGP, CDOC and biopolymers. While, for SDI and $MFI_{-0.45}$ measurement, the samples were collected in 30 L autoclaving bag (Sterilin, USA).

One media filter was selected from the first stage of DMF for monitoring purposes. The filtrate of the selected filter was monitored over time (before and after backwashing). Microbial ATP, BGP, SDI and $MFI_{-0.45}$ were monitored.

Table 6.2: The water properties of the SWRO influent and effluent (potable water).

Parameter	Influent	Potable water
pH	8.0 - 8.2	8.0-8.3
Conductivity	54 mS/cm	90-400 µS/cm
Temperature	30-34 °C	32-35 °C

6.2.3 Comparing DMF pre-treatment to DAF-DMF pre-treatment

The removal of fouling potential through the studied full-scale SWRO plant (plant A, Fig. 6.1) was compared with the removal of another full-scale SWRO desalination plant (plant B, Fig 6.2), in which the pre-treatment (plant B) included dissolved air floatation (DAF) with 1 - 5 mg/L of Fe^{3+} and two stage DMFs (with $0.3 - 1.5$ mg/L of Fe^{3+}). The raw seawater of both plant are different.in terms of the location and water quality. It is worth mentioning that the design and operating parameters (including, type, bed height, size of media, filtration rate, contact time, etc) in the DMFs (both stages) in plant A and plant B are identical except the fact that in the DMFs of plant B are backwashed with SWRO brine water.

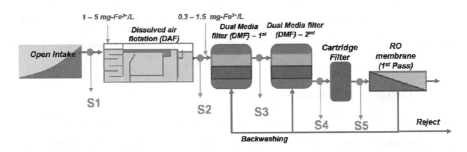

Figure 6.2: The SWRO pre-treatment scheme of Plant B.

6.2.4 Water quality characteristics
- ### Microbial ATP

Microbial ATP was measured along the pre-treatment of the SWRO using the ATP filtration method which is described in Abushaban et al. (2019b). Shortly, (i) a seawater sample was filtered through sterile 0.1 μm PVDF membrane filters. (ii) The retained microorganisms on the membrane filter surface were rinsed with 2 mL of sterilized artificial seawater water. (iii) 5 mL of Water-Glo lysis reagent (Promega Corp., USA) was passed through the filter to extract the microbial ATP from the retained cells. (iv) ATP of the filtrate was measured by mixing 100 μL aliquot with 100 μL of ATP Water-Glo detection reagent. The average emitted light measured by the luminometer (GloMax®-20/20, Promega Corp.) was converted to microbial ATP concentration based

145

on a calibration curve. Microbial ATP was measured onsite and all samples were collected and measured in triplicate measurement. In total, 88 samples were collected for ATP analysis.

▪ Bacterial growth potential (BGP) measurement

BGP measurement indicates the ability of bacteria to grow using the nutrients available in a seawater sample. BGP was measured using the method described in Abushaban et al. (2019a). Seawater samples were pasteurised in the plant's laboratory for 30 min to inactivate microorganisms and thereafter shipped to IHE-Delft, the Netherlands for analysis. Each pasteurized sample was distributed in triplicate into 30 mL carbon-free vials and each vial was inoculated with 10,000 cells/mL (intact cell concentration measured by flow cytometry) using a native/indigenous microbial consortium from the raw seawater. Seawater samples were incubated at 30 °C and bacterial growth was monitored using microbial ATP in seawater for 5 days.

▪ Organic fractions

Biopolymers and hydrophilic dissolved organic carbon (CDOC) was detected using Liquid chromatography - Organic Carbon Detection (LC-OCD). Seawater samples were shipped in a cool box (5 °C) to DOC-Labor Dr.Huber lab in Germany for analysis.

▪ SDI and MFI$_{-0.45}$

The two ASTM methods for particulate fouling potential in RO system were used (namely; SDI and MFI$_{0.45}$). SDI and MFI$_{-0.45}$ were measured using a portable SDI /MFI Analyzer (Convergence, Netherlands). It should be noted that the reported value should not exceed 75 % of the maximum value (5 %/min) (ASTM 2002). In case of high particulate fouling potential, shorter time needs to be used for the SDI test such as 10 min (SDI$_{10}$) or 5 min (SDI$_5$). If the reported value exceeds 75 % of SDI$_5$ (maximum value = 15 %/min), another test should be used such as MFI$_{0.45}$ (ASTM 2002). For this purpose, SDI$_5$ was measured in the seawater intake and SDI$_{15}$ was measured along the pre-treatment (after DMF1, DMF2 and CF).

6.3 RESULTS AND DISCUSSION

6.3.1 Seawater intake water quality

High variation in the seawater quality at the intake was observed (Table 6.3). The highest chlorophyll a concentration, microbial ATP concentration and $MFI_{0.45}$ were observed on day 3. The high concentration of chlorophyll a may indicate higher algal concentration in the seawater intake comparing on day 3 compared with other days. The highest SDI_5 and turbidity were measured on the day that chlorination was performed (4th day).

Table 6.3: seawater properties in the intake.

Date	Chlorophyll a µg/L	Microbial ATP (ng-ATP/L)	Turbidity (NTU)	SDI_5 (%)	$MFI_{0.45}$	Notes
Day 1	10.3	NA	1.1	> 15	14.2	
Day 2	17.1	240	1.1	> 15	21.0	
Day 3	32.5	535	0.9	> 15	22.4	
Day 4	6.9	27	2.8	> 15	16.5	Chlorination (1mg/L)

NA: data is not available.

6.3.2 Particulate parameters

- **Silt density index (SDI)**

SDI_5 in the seawater intake was more than the maximum reported value ($SDI_5 = 15$), based on ASTM method (ASTM 2002). The high SDI_5 indicates that the seawater (at the intake) had a high particulate fouling potential. Salinas Rodriguez et al. (2019) also reported high SDI values in raw seawater (North Sea) ranging between 9 and 28 %/min. Significant removal of particles was observed through the first stage of DMF, where 0.7 – 1.7 mg-Fe^{3+}/L were added (Fig. 6.2a) and the SDI_{15} after the first stage of DMF averaged 5 %/min. The measured SDI_{15} after the first stage of DMF was higher than the reported values by Bonnelye et al. (2004), who reported SDI_{15} below 3.3 %/min after DMF with 1 mg-Fe^{3+}/L as inline coagulation. Some literatures reported even higher SDI_{15} (> 6.6 %/min) after DMF (Sabiri et al. 2017). Further improvement in SDI_{15} was noted through the second stage of DMF and cartridge filtration with an average SDI_{15} of 3.4, and 3.1, respectively. Higher SDI_{15} after the second stage of DMF on day 4 ($SDI_{15} = 5$) was measured and this could be attributed to the de-attachment of a biofilm layer present on the media of the second stage of DMF when chlorine is dosed. This was not seen after the first stage of DMF as the backwashing frequency of the first stage of DMF is much higher (daily) compared with the backwashing frequency of the second stage of DMF (> 5 days). Thus, biofilm formation in the second stage of DMF is expected to be greater because of lower backwash frequency. It should be noted that the measured SDI_{15} in the SWRO feed water (after cartridge filtration) meets the guarantee level of the manufacturer ($SDI_{15} < 4$).

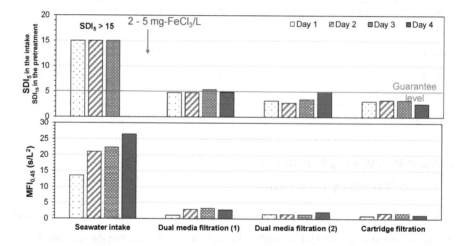

Figure 6.3: (a) Silt density index of 5 minutes (seawater intake) and 15 minutes and (b) Modified fouling index of 0.45 µm through the SWRO pre-treatment processes of the SWRO desalination plant. Chlorination was applied on Day 4.

- **Modified fouling index (MFI-0.45)**

The $MFI_{0.45}$ of the seawater (intake) ranged between 14 and 26.5 s/L^2 with an average of 21 s/L^2 (Fig. 6.2b). This is lower than the $MFI_{0.45}$ in the North Sea (20 - 310 s/L2) reported by Salinas Rodriguez et al. (2019). High $MFI_{0.45}$ values were measured on day 2 to day

4. The high $MFI_{0.45}$ values of day 2 and day 3 could be due to the high concentration of algae on these days, in which high chlorophyll a concentrations were observed 17.1 and 32.5 µg/L, respectively. Although lower concentrations of chlorophyll a were measured in the seawater intake on day 4, the $MFI_{0.45}$ values were quite high (26.5 and 21.7 s/L^2). This could be attributed to chlorination of the seawater source. Chlorination breaks down organic matter (mainly algae) into small fractions (Weinrich 2015), which may increase $MFI_{0.45}$.

A significant reduction (88 %) in $MFI_{0.45}$ was noticed through the first stage of DMF with the inline coagulation dosage (0.7 – 1.7 mg-Fe^{3+}/L). The $MFI_{0.45}$ declined from 21 to 2.5 s/L^2 (Fig. 6.2b). This is an extremely low of $MFI_{0.45}$ compared to the reported values by Salinas Rodriguez et al. (2019) after DMF (12 – 170 s/L^2). The high removal of $MFI_{0.45}$ in the first stage of DMF was confirmed by SDI measurements. Further reduction in $MFI_{0.45}$ was achieved through the second stage of DMF and cartridge filtration. $MFI_{0.45}$ ranged from 1.4 to 2.1 s/L^2 and from 0.9 to 1.7 s/L^2 after the second stage of DMF and after CF, respectively. This is consistent with SDI_{15} measurements. However, the advantage of $MFI_{0.45}$ over the SDI is that it has no maximum limit and it is based on a filtration mechanism i.e. cake filtration. Overall, high removal of particulate fouling potential was measured through the pre-treatment of the SWRO plant, in which a decrease in $MFI_{0.45}$ from 21 to 1.4 s/L^2 was achieved.

6.3.3 Biomass quantification

Microbial ATP in the seawater (intake) varied from 27 to 525 ng-ATP/L (Fig. 6.3), and the lowest microbial ATP concentration was observed on the fourth day when chlorination was applied in the intake pipe. The highest microbial ATP concentration was observed on day 3, which could be attributed to the algal ATP as high chlorophyll A concentration was also found on the same day. Another reason to explain the high microbial ATP on day 3 may be that carbon release from algae (i.e. TEP) led to higher bacterial growth and thus higher microbial ATP. The measured microbial ATP in the raw seawater is within the range reported in the North Sea (20 - 1,000 ng-ATP/L) by Abushaban et al. (2018).

Microbial ATP declined significantly (85 %) from 385 ng-ATP/L in the intake to less than 60 ng-ATP/L after the first stage of DMF. Abushaban et al. (2018) reported 67 % removal of microbial ATP through conventional pre-treatment (coagulation, flocculation and filtration) in a full-scale SWRO desalination plant with higher coagulant dose (3.8 mg-Fe^{3+}/L) than the applied here (0.7 – 1.7 mg-Fe^{3+}/L). When chlorination was applied, higher microbial ATP levels were measured after the first stage of DMF compared to the microbial ATP in the seawater intake and after the first stage of DMF in the absence of chlorine. This might be due to the breakdown of the biofilm present on the media or insufficient residual chlorine concentration (that reached the first stage of DMF) as only 1 mg Cl_2/L was applied into the seawater intake. Further reduction in microbial ATP was observed through the second stage of DMF, but no further reduction was observed through the cartridge filtration. Microbial ATP concentrations ranged between 12 and 22 ng-ATP/L after the second stage of DMF and between 10 and 21 ng-ATP/L after cartridge filtration. The insignificant reduction in microbial ATP through cartridge filtration is due to the short retention time compared to DMF, as well as the filter pore size (5 μm).

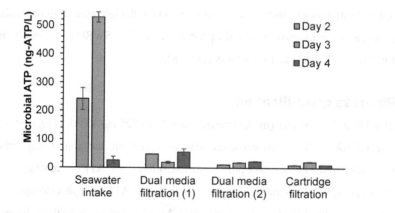

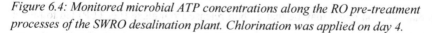

Figure 6.4: Monitored microbial ATP concentrations along the RO pre-treatment processes of the SWRO desalination plant. Chlorination was applied on day 4.

On average, the removal of microorganisms through the SWRO pre-treatment was high (95 %), and microbial ATP decreased from 385 ng-ATP/L in the seawater intake to 14 ng-ATP/L in the SWRO feed water. The highest removed was observed in the first stage of DMF with inline coagulation. The measured microbial ATP concentration in SWRO feed is equivalent to 16,000 intact cells/mL (using the correlation reported in Chapter 3) which is 1.6 times the concentration of cells used to inoculate BGP samples, suggesting that significant bacterial growth could occur in a SWRO system fed with feed water that can support this level of microorganisms.

6.3.4 Biological/organic fouling parameters

- **Organic fractions**

CDOC and biopolymer concentration were measured on day 2 (Fig. 6.4). The measured CDOC concentration in the seawater intake was 1.5 mg-C/L, which decreased 16.5 % through the first stage of DMF to 1.3 mg-C/L and another 16.5 % through the second stage of DMF to 1.1 mg-C/L. The observed removal of CDOC in DMFs is close to the observed removal (0.15 mg-C/L, 15 %) in the pressurized DMF (with $0.3 - 1.5$ mg-Fe^{3+}/L) of plant B in Chapter 4. Higher CDOC concentration was observed after the cartridge filter which could be due to the addition of antiscalant (Vrouwenvelder et al. 2000). A similar result was also reported by Jeong et al. (2016). Another possible reason for the higher concentration of organics after cartridge filter is the de-attachment of algae from the cartridge filter as higher SDI and MFI values were also measured on day 2.

A similar trend was observed for the biopolymer concentration, knowing that the percentage of biopolymers is approximately 10 % of the measured CDOC concentration. The observed removal of biopolymers in the first stage of DMF (72 µg-C/L, 41 %) is double the reported removal (35 µg-C/L, 29 %) in plant B (Chapter 4).

Overall, the achieved removal of CDOC and biopolymers along the SWRO pre-treatment processes were 24 % and 37 %, respectively, in which the highest removal was found in the first stage of DMF with inline coagulation ($0.7 - 1.7$ mg-Fe^{3+}/L).

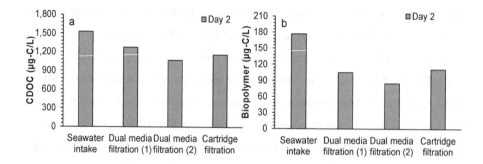

Figure 6.5: Monitored hydrophilic dissolved organic carbon (CDOC) and biopolymer concentrations along the RO pre-treatment processes of the SWRO desalination plant, as measured by LC-OCD.

▪ Bacterial Growth Potential

The average BGP in the seawater (intake) was 320 μg-C/L as glucose (Fig. 6.5). Only slight removal of BGP was observed in the first stage of DMF where $0.7 - 1.7$ mg-Fe^{3+}/L is added (70 μg-C/L as glucose, 22 %) and in the second stage of DMF (20 μg-C/L as glucose, 8 %). The observed removal of BGP in both stages of DMF is considerably lower than the reported BGP removal in two full-scale desalination plants described in Chapter 4 (Abushaban et al. 2019a), where the removal was 55 % (190 μg-C/L as glucose) in a gravity DMF coupled with 2.3 mg/L $FeCl_3$ and 68 % (156 μg-C/L as glucose) in pressurized DMF with 13.6 mg/L $Fe_2(SO_4)_3$. Moreover, Weinrich et al. (2011) reported 23 - 80 % removal of AOC ($40 - 280$ μg-C/L as acetate) through the sand filtration step in the Tampa bay desalination plant. The poor removal of BGP through the first and second stages of DMF in this study resulted in SWRO feed water with a high average concentration of BGP (190 μg-C/L as glucose), despite extensive pre-treatment. The level of BGP achieved after the pre-treatment (190 μg-C/L as glucose) is close to the BGP (200 μg-C/L as glucose) in the SWRO feed water in plant C (described in Chapter 4) where the cleaning-in-place frequency was 6 CIP's/year. The overall BGP removal (41 %) is consistent with the observed removal of biopolymers (37.3 %), which suggests that the removal of organics in SWRO pre-treatment systems is limited compared with the removal of particulate fouling (93 %, as MFI) and biomass (95 %, as microbial ATP).

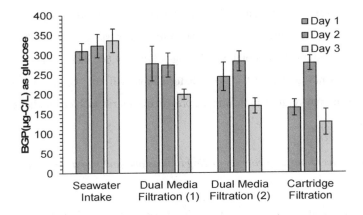

Figure 6.6: Bacterial growth potential along the RO pre-treatment processes of a full-scale SWRO desalination plant.

6.3.5 Removal efficiency of fouling potential in DMF pre-treatment

In Table 6.4, a summary of the measured removal of various fouling indicators and parameters through the pre-treatment processes of a full-scale SWRO desalination plant. It can clearly be seen that higher removal rates (> 80 %) were achieved for particulate fouling indices compared with organic/biological fouling parameters, in which the highest removal was obtained in the first stage of DMF with inline coagulation (0.7 – 1.7 mg-Fe^{3+}/L). The second stage DMF followed by cartridge filtration further improved the seawater quality in terms of particulate fouling potential. The measured SDI_{15} in the SWRO feed water (3.1 %/min) is below the recommended value of the membrane manufacture. Moreover, using the prediction model for particulate fouling in RO systems by Salinas Rodriguez et al. (2019) and Boerlage et al. (2003), the $MFI_{0.45}$ values measured in the SWRO feed water (1.4 s/L^2), particulate fouling will not occur in the SWRO system within a 2 year period (see figure S6.1). Therefore, it can be concluded that two stage dual media filtration pre-treatment in SWRO showed very good removal of particulate fouling potential.

Furthermore, high removal of microbial ATP was also observed (97%) through the SWRO pre-treatment, with the highest removal reported in the first stage DMF (85 %). However, the low concentration of microbial ATP (14 ng-ATP/L ~ 1,600 intact cells/mL) in the SWRO feed does not guarantee that biofilm formation will not occur. On the contrary, microorganisms may re-grow rapidly in the presence of nutrients/ biodegradable organic compounds.

Table 6.4: Summary of the removal of various fouling indices and fouling parameters in SWRO pre-treatment.

	Removal in DMF1	Removal in DMF2	Removal in CF	Overall removal
SDI, %/min (%)	>10 (> 65 %)	1.6 (32 %)	0.3 (9 %)	>11.9 (>80 %)
$MFI_{0.45}$, s/L^2 (%)	18.5 (88 %)	0.9 (36 %)	0.2 (12 %)	19.6 (94 %)
Microbial ATP, ng-ATP/L (%)	325 (85 %)	43 (72 %)	3 (18 %)	371 (97 %)
BGP, µg-C/L (%)	70 (22 %)	20 (8 %)	40 (17 %)	130 (41 %)
CDOC, µg-C/L (%)	252 (17 %)	209 (17 %)	-93 (-9 %)	368 (24 %)
Biopolymers, µg-C/L (%)	72 (41 %)	20 (24 %)	-26 (-31 %)	66 (37 %)

Lower removal (24 - 41 %) of organic and biological fouling potential was observed through the SWRO pre-treatment compared with the observed removal of particulate fouling and microorganisms (as microbial ATP). Similar to the particulate fouling potential and microbial ATP, the first stage DMF showed the highest removal of organic and biological fouling potential including 16.5 % of CDOC (252 µg-C/L), 41 % of biopolymers (72 µg-C/L) and 22 % of BGP (70 µg-C/L). The low removal of organic and biological fouling potential through the SWRO pre-treatment system could be attributed to a low level of biological activity in the media filters. This could be caused by operational practice such as (i) frequent chlorination of the intake and dechlorination after the media filtration units, (ii) too short empty bed contact times in the filter media or (iii) applying backwashing conditions that remove/wash out biofilm layers on filter media, which are required to degrade organic matter (nutrients) preventing further bacterial growth in the downstream SWRO membranes.

▪ Monitoring DMF 1st stage over time

For further investigation, the filtrate of one first-stage DMF was selected for in-depth monitoring (before and after backwashing), including SDI$_5$, MFI$_{0.45}$, microbial ATP and BGP (Fig. 6.6).

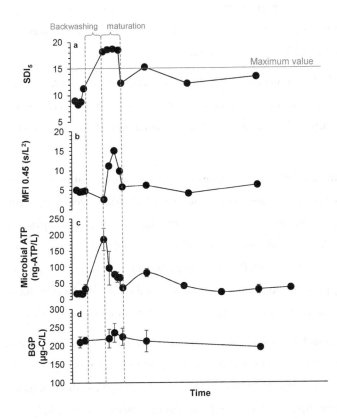

Figure 6.7: MFI$_{0.45}$, SDI$_5$, microbial ATP and Bacterial Growth Potential (BGP) measured before and after backwashing in the first stage of a dual media filter.

SDI_5 and $MFI_{0.45}$ before backwashing were approximately 8 %/min and 5 s/L^2, respectively (Fig. 6.6a, 6.6b). High values of SDI_5 (> 15 %/min) and $MFI_{0.45}$ (> 17 s/L^2) were measured during the maturation phase, which started immediately after backwashing and had a duration of 30 minutes. The high SDI_5 and $MFI_{0.45}$ values during maturation indicate significant removal of colloidal particles. However, the SDI_5 (12 - 15 %/min) and $MFI_{0.45}$ (5 – 7 s/L^2) values measured after maturation are high. This may suggest that the backwashing protocol needs to be optimized and/or the maturation time needs to be prolonged. The high SDI_5 and $MFI_{-0.45}$ values observed during maturation were not observed when the pre-treatment processes were monitored (presented in Fig. 6.2) as only one filter (out of 24 in total) is backwashed at any point in time and the (mixed) samples were collected from the main header.

The trends in the microbial ATP concentration (Fig 6.6c) was typically similar to the observed trend of SDI_5 and $MFI_{-0.45}$. However, the measured microbial ATP after maturation was not significantly higher than the measured microbial ATP before backwashing.

Insignificant change in the BGP was found before and after backwashing of the first stage of DMF (Fig. 6.6d) which could indicate either the absence of a biofilm layer on the media filter or the biofilm layer attached on the media was not influenced by backwashing (strong biofilm layer) which is unlikely as low removal of BGP was detected through the first stage of DMF (Fig. 6.6). Low reduction in BGP was observed over time (10 μg-C/L as glucose within > 6 h) indicating poor development of the biofilm on the filter media (Fig. 6.6d). This suggests that the media filter functions like a sieve and not a biologically active filter.

Three options were suggested to explain the low biological activity in the first stage of DMF (i) too harsh backwashing conditions that completely remove the biologically active layer, (ii) frequent chlorination which destroys the biofilm and/or (iii) too short an EBCT for biodegradation to occur. Applying too harsh backwashing conditions is not likely to be the main reason for the low BGP removal as high values of $MFI_{-0.45}$ and SDI_5 were still measured after maturation of the first stage of DMF suggesting that the duration of backwashing should be extended in order to improve filterability in subsequent filtration cycles. Moreover, a too short EBCT is not likely to be the culprit either as higher removal of both BGP and organic matter were measured in another DMF pre-treatment system in the full-scale SWRO plant described in Chapter 5 of this thesis, with similar EBCT. Consequently, frequent chlorination of the pre-treatment is believed to be the main reason for the loss in activity of the biofilm layer, as chlorination is applied weekly in the seawater intake (Li et al. 2017b, De Vera et al. 2019). Consequently, free chlorine reaches the media filtration, as can be seen based on microbial ATP monitoring (Fig. 6.4), and destroys the biofilm layers on the filter media. It should be noted that biofilms need long time to form/develop on the filter media (2 – 3 days, see chapter 4 of this thesis, Fig S4.2). Thus, in order to further reduce the Bacterial Growth Potential of RO feed water, priority should be given to (i) reducing the frequency of chlorination in intakes and/or (ii) performing the neutralization step *before* media filtration, and not *after* as is current practice in many full-scale SWRO plants. Neutralizing chlorine before media filtration will support the development of an active biofilm layer on the media, capable of degrading/removing easily biodegradable organic matter in the feed water. Additionally, longer EBCT's may also significantly improve the biodegradation of organic matter as well as the removal of biological fouling potential in media filtration.

6.3.6 Comparing DMF and DAF-DMF pre-treatment

The overall removal achieved through the SWRO pre-treatment (inline coagulation with $0.7 - 1.7$ mg-Fe^{3+}/L, two stages of DMF and cartridge filtration) in plant A was compared to the removal with DAF coupled with $1 - 5$ mg-Fe^{3+}/L, inline coagulation with $0.3 - 1.5$ mg- Fe^{3+}/L, two stages of DMF and cartridge filtration (plant B) (Table 6.5). It should be noted that the design and operating parameters (including, the type, high and size of media, filtration rate contact time, etc) of the DMFs (both stages) in plant A and plant B are similar except that the DMFs in plant B are backwashed with SWRO brine and not filtered seawater as in the case in plant A.

A similar overall removal of particulate fouling potential was achieved in plant A and B. However, the particulate fouling potential measured in plant B (40.5 s/L^2) was considerably higher than in plant A (21 s/L^2). It can be seen that the highest removal of particulate fouling potential was observed in the first stage of DMF in plant A and in the DAF + DMF (first stage) in plant B. The SDI and MFI after the first stage of DMF in plant A and plant B were in the same range, which suggests that DAF in plant B did not contribute significantly to the removal of particulate fouling potential. However, DAF could reduce the particles loading to the succeeding DMF and thus increase the filtration rate of DMF. Kim et al. (2011) reported comparable SDI values in a pilot SWRO DAF system (5.7 %/min of SDI_{15} after DMF and 4.7 %/min when DAF was coupled with DMF).

Microbial ATP concentrations in the raw seawater of plant A and plant B were in the same range (Table 6.5). Even though high removal (48 %) of microbial ATP was observed in the DAF of plant B, the overall removal of microbial ATP through the pre-treatment of both plants were similar (371 and 335 ng-ATP/L). This may indicate that the addition of DAF did not increase the removal of microbial ATP.

The biological/organic fouling potential observed in the raw seawater of plant B comparing to plant A, with approximately 50 % higher in terms of BGP, CDOC and biopolymer concentration than in plant A. Moreover, the actual removal of biological/organic fouling potential in plant B (368 µg/L of BGP, 449 µg/L of CDOC and 163 µg/L of biopolymers) was significantly higher than in plant A (130 µg/L of BGP, 368 µg/L of CDOC and 66 µg/L of biopolymers). The additional removal in plant B was mainly attributed to the DAF system, where 260 µg-C/L of BGP, 109 µg-C/L of CDOC and 83 µg-C/L of biopolymer were removed due to the higher applied coagulant dose (1 - 5 mg-Fe^{3+}/L).

To summarize, comparing DMF pre-treatment with DAF-DMF pre-treatment showed that DAF significantly improved the removal of biological/organic fouling potential in terms of BGP and biopolymer (Table 6.5). However, the additional removal of both particulate fouling potential and microbial ATP in the DAF system was insignificant. Nevertheless, DAF may improve operation of the DMF by decreasing head loss and improving filter backwashing.

Table 6.5: Comparing DMF pre-treatment (plant A) to DAF-DMF pre-treatment (plant B).

Parameter	Plant	Raw seawater	DAF	DMF1	DMF2	CF	Overall removal
SDI, %/min (% removal)	A	>15*	-	5 (> 65 %)	3.4 (32 %)	3.1 (9 %)	>11.9 (>80 %)
	B	>15*	NA	4.8 (> 68 %)	3.9 (19 %)	3.5 (10 %)	>11.5 (>76 %)
$MFI_{0.45}$, s/L^2 (% removal)	A	21	-	2.5 (88 %)	1.6 (36 %)	1.4 (12 %)	19.6 (93.5 %)
	B	40.5	NA	3.6 (91 %)	1.7 (52 %)	1.3 (24 %)	39.2 (97 %)
Microbial ATP, ng-ATP/L (% removal)	A	385	-	60 (85 %)	17 (72 %)	14 (18 %)	371 (97 %)
	B	370	191 (48 %)	85.5 (55 %)	42.5 (50 %)	35 (18 %)	335 (91 %)
BGP, µg-C/L (% removal)	A	320	-	250 (22 %)	230 (8 %)	190 (17 %)	130 (41 %)
	B	460	200 (57 %)	120 (40 %)	107 (11 %)	92 (14 %)	368 (80 %)
CDOC, µg-C/L (% removal)	A	1528	-	1276 (17 %)	1067 (17 %)	1160 (-9 %)	368 (24 %)
	B	2015	1904 (6 %)	1588 (11 %)	1590 (0 %)	1566 (1.5 %)	449 (22 %)
Biopolymers, µg-C/L (% removal)	A	177	-	105 (41 %)	85 (24 %)	111 (-31 %)	66 (37 %)
	B	311	228 (27 %)	194 (15 %)	151 (22 %)	148 (2 %)	163 (53 %)

*measured as SDI_5

159

6.4 CONCLUSIONS

- The removal of particulate fouling potential (SDI, MFI$_{0.45}$), biological/organic fouling potential (BGP, CDOC and biopolymer concentration) as well as microbial ATP was monitored through the pre-treatment (two stage DMF with 0.7 – 1.7 mg-Fe^{3+}/L) of a full-scale SWRO desalination plant.

- More than > 85 % removal of particulate fouling potential (in terms of SDI, MFI$_{0.45}$) and microbial ATP was achieved through pre-treatment (inline coagulation and two stage DMFs), in which the highest removal (65 – 85 %) was observed in the first stage of DMF. However, significantly lower removal of the organic/biofouling potential in terms of BGP, CDOC and biopolymers was achieved through the first stage of DMF (22 – 41 %) and after pre-treatment (24 – 41 %). This was attributed to frequent (weekly) chlorination of the intake and the fact that de-chlorination takes place *after* media filtration, resulting in damage to the biofilm layer on the filter media, which was also reflected in the ATP and BGP measurements.

- The removal of biological/organic fouling potential observed in two stage DMFs in another SWRO plant (two stage DMFs with 0.3 – 1.5 mg- Fe^{3+}/L preceded by dissolved air flotation (DAF-DMF)) was similar to the observed removal in two stage DMFs coupled with inline coagulation. The low removal of biological/organic fouling potential in the DMFs of the second SWRO plant was attributed to the use of SWRO brine to backwash the DMF, which may have hindered the formation of biofilm after backwashing due to the osmotic shock experienced by by the bacteria. This was confirmed by ATP and BGP measurements during and immediately after brine backwashing.

- The application of the newly developed ATP and BGP methods showed that they can be used to improve current practices in SWRO such as chlorination/dechlorination as well as backwashing, in order to optimize the removal of biological/organic fouling potential through the pre-treatment.

- The overall removal achieved through two stages of DMFs (with 0.7 – 1.7 mg-Fe^{3+}/L) was compared to DAF (with 1 – 5 mg- Fe^{3+}/L) and two stage DMF (with 0.3 – 1.5 mg- Fe^{3+}/L). It was found that DAF significantly improved the removal of biological/organic fouling potential, with the removal of BGP and biopolymers increasing to 80 % and 53 %, respectively.

6.5 ANNEXES

Annex A: Operating time of RO membrane system based on MFI in RO feed water

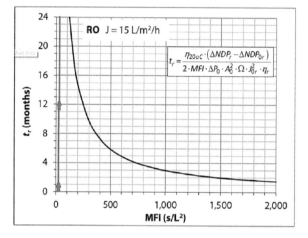

Figure S6.1: The expected time of particulate fouling in RO membrane system calculated based on MFI value in the RO feed water (Salinas Rodriguez et al. 2019).

7

CONCLUSIONS AND FUTURE PERSPECTIVE

7.1 CONCLUSIONS

In this research, a method to quantify microbial ATP in seawater was developed using new reagents specially developed for seawater. The new ATP method was subsequently used to monitor bacterial growth potential (BGP) in seawater using an indigenous microbial consortium. Afterwards, the ATP-based BGP method was applied to assess the pre-treatment of several full-scale SWRO desalination plants. Finally, the correlation between BGP of SWRO feed water and the operational performance of SWRO membrane systems was investigated.

7.1.1 ATP measurement in seawater

The direct addition of newly developed reagents (for ATP extraction and detection) to seawater (ATP-direct method) allowed the detection of free, total and microbial ATP at very low levels (LOD = 0.3 ng-ATP/L) in seawater. However, pH changes and the presence of iron in seawater, negatively affected the luminescence signal. These effects can be taken into consideration by establishing a calibration line considering the seawater matrix (Fe, pH). However, seawater characteristics change considerably along pre-treatment processes, requiring the preparation of several different calibration lines to correct these effects. Thus, the ATP-direct method is more suitable for monitoring raw seawater where frequent changes in the seawater matrix are not expected. To overcome the issue of seawater matrix/quality changes (in terms of iron concentration and pH) during pre-treatment, a filtration process (using 0.1 µm filtration) was introduced (ATP-filtration method) to capture marine microorganisms on a membrane surface, eliminating interference of the seawater matrix with ATP measurement. Moreover, the use of filtration allowed direct determination of microbial ATP, instead of measuring both total and free ATP. The ATP-filtration method is fast (<5 min), reproducible (VC = 7 %) and sensitive (LOD = 0.06 ng-ATP/L, equivalent to 70 cells/mL). The LOD's of the two ATP methods developed were in the same range as the reported LOD of ATP measurement in freshwater ATP (0.05 - 1 ng-ATP/L). Microbial ATP concentration measured using the ATP-filtration method correlated with intact cell concentration measured by flow cytometry ($R^2 = 0.72$, Rho = 0.88, $P \ll 0.001$, n = 100). The average microbial ATP concentration per marine bacterial cell is $ca.$ 8.59×10^{-7} ng-ATP/cell.

Both the ATP-direct and ATP- filtration method were applied to monitor microbial ATP along the pre-treatment train of SWRO desalination plants and significant reduction (> 50 %) was found after dual media filtration (DMF) combined with 1.3 - 4.5 mg-Fe^{3+}/L coagulant dose. The measured microbial ATP concentration using the ATP-filtration method was comparable (± 5 % difference, R^2 = 0.95, n = 125) to the concentration measured using the ATP-direct method. Moreover, the use of the newly developed ATP methods to monitor the performance of media filtration provided useful information in terms of backwashing duration, maturation, biodegradation and biofilm formation on media filters, suggesting that ATP can be a useful tool to optimize biofilm formation and biodegradation in media filters, with the aim to further reduce the BGP of SWRO feed water. However, it can be also be applied for applications ballast water, swimming pool water, cooling towers, food industry or in freshwater where very low LOD is required.

7.1.2 Bacterial growth potential measurement in seawater using an indigenous microbial consortium

The newly developed ATP methods were used to measure BGP employing an indigenous microbial consortium. The method comprises four main steps including bacterial inactivation, bacterial inoculation, incubation, and bacterial growth monitoring based on microbial ATP measurement. The bacterial growth of an indigenous microbial consortium was affected by the incubation temperature, with the highest bacterial growth achieved at an incubation temperature similar to the original temperature of the inoculum, which is a limitation of using an indigenous microbial consortium. However, this effect was overcome by using a calibration line (BGP as a function of the carbon concentrations) at a constant incubation temperature for each seawater location, which allowed the bacterial yield of different seawater samples at different locations to be compared. The bacterial yield in seawater was measured at five different locations globally (Tasman Sea, Arabian Sea, Persian Gulf, Gulf of Oman and North Sea) and relatively low variations (1 - 1.5 ng-ATP/µg C-glucose) were observed between the different locations.

The LOD of the developed BGP method is 13 µg C/L expressed as glucose equivalents, in which artificial seawater was used as a blank. Further development to lower the LOD (~1 µg C-glucose) is still required for low nutrient seawater and low seawater

165

temperatures. In this research, the lowest BGP measured in SWRO feed water was five times higher than the LOD of the method. At this BGP concentration, the SWRO membranes were cleaned in place (CIP) every 3 years. Lowering the LOD of the BGP method could broaden the applicability of the method as well to different types of water, for instance brackish ground water.

The ATP-based BGP method was used to monitor BGP in the North Sea over a period of 12 months. The BGP ranged between 45 µg C-glucose/L in the winter and 385 µg C-glucose/L in the autumn. The current BGP method is time consuming, tedious and prone to manual errors and is not suitable for routine measurement in full-scale SWRO plants. The development of an online tool/analyser to perform routine BGP analyses is highly desirable. However, many issues need to be solved including the bacterial inactivation protocol, LOD, contamination from the materials in contact with the seawater, cleaning protocol during use, and duration of the test.

7.1.3 Assessing SWRO pre-treatment performance in full-scale desalination plants

In this research, the pre-treatment trains of five full-scale SWRO plants and one pilot scale plant were monitored. All six plants had open intakes and different pre-treatment processes. Table 7.1 summarizes the removal of BGP through the six SWRO plants. The highest BGP removal (40 – 55 %) was observed in DMFs in combination with inline coagulation. Higher removal of BGP was achieved in media filtration compared to ultrafiltration (10 – 33 %) which could be attributed to the biodegradation/adsorption in media filtration, due to the longer contact time in media filters (4 - 5 min) compared with UF (< 10 sec) as well as the addition of coagulation.

Media filtration pre-treatment in SWRO removed a significant amount of the biological/organic fouling potential, with more than 50 % removal of BGP in pressurised and gravity DMF's combined with inline coagulation in two different desalination plants. However, much lower BGP removal was observed in the DMFs in two other SWRO plants, due to frequent chlorination of the intake (with dechlorination *after* media filtration) and filter backwashing with SWRO brine, both of which may stress/destroy the active biofilm layer on the filter media. This was confirmed by ATP measurements made during chlorination and after brine backwashing. Neutralizing chlorine *before* media filtration will support the development of an active biofilm layer on the media, capable of degrading/removing easily biodegradable organic matter in the feed water. Additionally, backwashing the filters with filtered seawater may also significantly improve the biological activity in the media filters as well as biodegradation of organic matter.

Table 7.1: Comparing BGP removal in different SWRO pre-treatment of six seawater plants

Plant	Plant scale	Pre-treatment	Coagulant dosage (mg-Fe^{+3}/L)	BGP in raw seawater	Overall BGP removal
1	Full-scale	DAF – Coag. – 2 stage DMF	1 – 5 in DAF and 0.3 - 1.5 as inline coagulation	460 µg/L	80 %
2	Full-scale	Coag. – 2 stage DMF	0.7 – 1.7 as inline coagulation	320 µg/L	40 %
3	Full-scale	DAF – UF	0.5 in DAF	400 µg/L	50 %
4	Full-scale	Coag. – DMF	0.8 as inline coagulation	350 µg/L	55 %
5	Full-scale	Coag. – Floc. – DMF	3.6 as inline coagulation	230 µg/L	72 %
6	Pilot scale	DMF – UF	0	330 µg/L	50 %

Dissolved air floatation (DAF) removed 20 – 50 % of the BGP, and the removal of BGP in DAF combined with ultrafiltration (UF) was comparable to the removal in DMF with inline coagulation (0.8 mg Fe^{3+}/L). DAF combined with two stage DMF in SWRO pre-treatment, achieved more than 80 % removal of BGP, which was roughly double the removal achieved in a two stage DMF. The high removal of biological/organic fouling potential by DAF was attributed to the high coagulant does applied (1 – 5 mg-Fe^{3+}/L). However, the addition of antiscalant increased the BGP concentration in SWRO feed water in two different SWRO plants, either due to contaminated storage vessels or make up water and/or the addition of antiscalant itself.

7.1.4 Correlating Bacterial Growth Potential of SWRO feed water to membrane performance

Investigating the relationship between the measured BGP in SWRO feed water and membrane performance is complicated by several factors: (i) several types of fouling may occur simultaneously in full-scale SWRO plants and (ii) the widespread intermittent use of non-oxidizing biocides to combat biofouling in full-scale SWRO facilities makes establishing any real correlation between the BGP of SWRO feed water and membrane performance very difficult. Moreover, to establish a robust correlation, a large number of SWRO desalination plants (in this study only five plants were investigated) in different parts of the world need to be monitored for longer periods of time with different operating conditions.

Despite these limitations, an attempt was made to investigate if any correlation existed between the measured BGP in SWRO feed water and fouling development over time. Based on data collected from four full-scale SWRO plants, a higher cleaning-in-place (CIP) frequency (used as a surrogate parameter for biofouling) corresponded to a higher BGP in the SWRO feed water. Moreover, the growth potential measured in SWRO feed water ($100 - 950$ µg-C/L) led to significant increase in the normalized pressure drop in the full-scale SWRO plants within a period of 3 months. These results are promising and may indicate the applicability of using BGP of SWRO feed water as a biological fouling indicator in SWRO systems. Accordingly, a safe level of BGP (<70 µg/L as glucose) is tentatively proposed for SWRO feed water in order to ensure a chemical cleaning frequency of once/year or lower. However, to establish a robust correlation and threshold level, many more SWRO plants need to be monitored at different locations for longer periods of time.

7.2   FUTURE PERSPECTIVE

While performing this research, several areas emerged where further research is still required. Bacterial growth potential in seawater, using an indigenous microbial consortium, was successfully measured based on the developed microbial ATP method. However, further development is still needed to lower the limit of detection and to accelerate bacterial growth for faster BGP detection. Although the current limit of detection of the BGP method was sufficient to measure BGP along SWRO pre-treatment trains of SWRO plants with open intakes. However, a much lower limit of detection may be required for SWRO pre-treatment with beach well pre-treatment or deep brackish ground water.

In this thesis, an attempt was made to determine a threshold BGP value for biofouling in SWRO feed water and to investigate if any correlation exists between the BGP of SWRO feed water and pressure drop development in full-scale SWRO plants. While this study comprised an evaluation of five full-scale SWRO plants, many more SWRO plants in different locations with a wide variety of seawater pre-treatment technologies need to be monitored to establish a robust correlation. The development of an online BGP analyser would make this task much easier.

Finally, SWRO pre-treatment systems studied in this research (Table 7.1) all significantly improved the particulate fouling potential of the feed water in the SWRO plants studied. However, high variations in BGP removal were observed, suggesting that more focus should be placed on optimizing the efficiency of SWRO pre-treatment, and in particular dual media filtration, to lower the biological/organic fouling potential of SWRO feed water and the impact of biofouling in SWRO membrane systems.

REFERENCES

Abd El Aleem, F. A., K. A. Al-Sugair, and M. I. Alahmad. 1998. Biofouling problems in membrane processes for water desalination and reuse in Saudi Arabia. International Biodeterioration and Biodegradation **41**:19-23.

Abushaban, A., M. N. Mangal, S. G. Salinas-Rodriguez, C. Nnebuo, S. Mondal, S. A. Goueli, J. C. Schippers, and M. D. Kennedy. 2018. Direct measurement of ATP in seawater and application of ATP to monitor bacterial growth potential in SWRO pre-treatment systems. Desalination and Water Treatment **99**:91-101.

Abushaban, A., S. G. Salinas-Rodriguez, N. Dhakal, J. C. Schippers, and M. D. Kennedy. 2019a. Assessing pretreatment and seawater reverse osmosis performance using an ATP-based bacterial growth potential method. Desalination **467**:210-218.

Abushaban, A., S. G. Salinas-Rodriguez, M. N. Mangal, S. Mondal, S. A. Goueli, A. Knezev, J. S. Vrouwenvelder, J. C. Schippers, and M. D. Kennedy. 2019b. ATP measurement in seawater reverse osmosis systems: Eliminating seawater matrix effects using a filtration-based method. Desalination **453**:1-9.

Aintablian, X. W. 2017. Water Desalination. Desalination Expands as Technology Becomes More Affordable. www.thoughtco.com.

Al-Ahmad, M., F. A. Abdul Aleem, A. Mutiri, and A. Ubaisy. 2000. Biofuoling in RO membrane systems Part 1: Fundamentals and control. Desalination **132**:173-179.

Al-Karaghouli, A., D. Renne, and L. L. Kazmerski. 2009. Solar and wind opportunities for water desalination in the Arab regions. Renewable and Sustainable Energy Reviews **13**:2397-2407.

Amy, G. L., S. G. Salinas Rodriguez, M. D. Kennedy, J. C. Schippers, S. Rapenne, P. J. Remize, C. Barbe, C. L. de O. Manes, N. J. West, P. L. Lebaron, D. Van der kooij, H. Veenendaal, G. Schaule, K. Petrowski, S. Huber, L. N. Sim, Y. Ye, V. Chen, and A. G. Fane. 2011. Water quality assessment tools Pages 3-32 Membrane-based desalination: an integrated approach (MEDINA). IWA Publishing.

Aoustin, E., A. I. Schäfer, A. G. Fane, and T. D. Waite. 2001. Ultrafiltration of natural organic matter. Separation and Purification Technology **22-23**:63-78.

ASTM. 2002. Standard test method for silt density index (SDI) of Water. Designation: D 4189 – 95 (Reapproved 2002). ASTM International, West Conshohocken, PA 19428-2959, United States.

ASTM D4189-14. 2014. Standard Test Method for Silt Density Index (SDI) of Water. ASTM International, West Conshohocken, PA (2014).

ASTM D8002 - 15. 2015. Standard Test Method for Modified Fouling Index (MFI-0.45) of Water. ASTM International, West Conshohocken, PA (2015).

ASTM Standard D 4012. 1981(Reapproved 2002). Standard test method for adenosine triphosphate (ATP) content of microorganisms in water. Book of standards (11.02), ASTM International.

Badruzzaman, M., N. Voutchkov, L. Weinrich, and J. G. Jacangelo. 2019. Selection of pretreatment technologies for seawater reverse osmosis plants: A review. Desalination **449**:78-91.

Bartram, J., J. Cotruvo, M. Exner, C. Fricker, and A. Glasmacher. 2003. Heterotrophic plate counts and drinking-water safety. IWA publishing.

Bendtsen, J., C. Lundsgaard, M. Middelboe, and D. Archer. 2002. Influence of bacterial uptake on deep-ocean dissolved organic carbon. Global Biogeochemical Cycles **16**.

Bergman, C., Y. Kashiwaya, and R. L. Veech. 2010. The Effect of pH and Free Mg2+ on ATP Linked Enzymes and the Calculation of Gibbs Free Energy of ATP Hydrolysis. The Journal of Physical Chemistry B **114**:16137-16146.

Berney, M., M. Vital, I. Hulshoff, H. U. Weilenmann, T. Egli, and F. Hammes. 2008. Rapid, cultivation-independent assessment of microbial viability in drinking water. Water Research **42**:4010-4018.

Boe-Hansen, R., H.-J. Albrechtsen, E. Arvin, and C. Jørgensen. 2002. Bulk water phase and biofilm growth in drinking water at low nutrient conditions. Water Research **36**:4477-4486.

Boerlage, S. F. E., M. Kennedy, M. P. Aniye, and J. C. Schippers. 2003. Applications of the MFI-UF to measure and predict particulate fouling in RO systems. Journal of Membrane Science **220**:97-116.

Bonnelye, V., M. A. Sanz, J.-P. Durand, L. Plasse, F. Gueguen, and P. Mazounie. 2004. Reverse osmosis on open intake seawater: pre-treatment strategy. Desalination **167**:191-200.

Bowman, F. W., M. P. Calhoun, and M. White. 1967. Microbiological methods for quality control of membrane filters. Journal of Pharmaceutical Sciences **56**:222-225.

Camper, A. 2001. Investigation of the biological stability of water in treatment plants and distribution systems. American Water Works Association.

Chia, F., and R. M. Warwick. 1969. Assimilation of Labelled Glucose from Seawater by Marine Nematodes. Nature **224**:720-721.

Cornelissen, E. R., L. Rebour, D. Van der Kooij, and L. P. Wessels. 2009. Optimization of air/water cleaning (AWC) in spiral wound elements. Desalination **236**:266-272.

De Vera, G. A., C. Lauderdale, C. L. Alito, J. Hooper, and E. C. Wert. 2019. Using upstream oxidants to minimize surface biofouling and improve hydraulic performance in GAC biofilters. Water Research **148**:526-534.

Denner, E. B., D. Vybiral, U. R. Fischer, B. Velimirov, and H. J. Busse. 2002. Vibrio calviensis sp. nov., a halophilic, facultatively oligotrophic 0.2 µm-filterable marine bacterium. International Journal of Systematic and Evolutionary Microbiology **52**:549-553.

DesalData. 2018. Worldwide desalination inventory (MS Excel Format), downloaded from DesalData.com. June 2016.

Dhakal, N. 2017. Controlling biofouling in seawater reverse osmosis membrane systems. Delft university of technology, Taylor & Francis Group.

Dietz, K., and A. Kulinkina. 2009. The Design of a Desalination Pretreatment System for Brackish Groundwater. Worcester polytechnic institute.

Dixon, M. B., T. Qiu, M. Blaikie, and C. Pelekani. 2012. The application of the bacterial regrowth potential method and flow cytometry for biofouling detection at the Penneshaw desalination plant in South Australia. Desalination **284**:245-252.

Edzwald, J. K., and J. Haarhoff. 2011. Seawater pretreatment for reverse osmosis: Chemistry, contaminants, and coagulation. Water Research **45**:5428-5440.

Escobar, I. C., and A. A. Randall. 2000. Sample storage impact on the assimilable organic carbon (AOC) bioassay. Water Research **34**:1680-1686.

Escobar, I. C., and A. A. Randall. 2001. Assimilable organic carbon (AOC) and biodegradable dissolved organic carbon (BDOC):: complementary measurements. Water Research **35**:4444-4454.

Eydal, H. S., and K. Pedersen. 2007. Use of an ATP assay to determine viable microbial biomass in Fennoscandian Shield groundwater from depths of 3-1000 m. Journal of Microbiological Methods **70**:363-373.

Farhat, N., F. Hammes, E. Prest, and J. Vrouwenvelder. 2018. A uniform bacterial growth potential assay for different water types. Water Research **142**:227-235.

Faujour, H., E. H. Koenig, C. Ventresque, C. de Vomecourt, M. Nicholson, Y. Ahmed, and R. van Leeuw. 2015. Fujairah 2 RO: Impact of effective seawater pre-treatment on RO membrane performance and replacement. World Congress on Desalination and Water Reuse 2015. The International Desalination Association, San Diego, CA, USA.

Ferrer, O., S. Casas, C. Galvañ, F. Lucena, A. Bosch, B. Galofré, J. Mesa, J. Jofre, and X. Bernat. 2015. Direct ultrafiltration performance and membrane integrity monitoring by microbiological analysis. Water Research **83**:121-131.

Flemming, H. C. 1997. Reverse osmosis membrane biofouling. Experimental Thermal and Fluid Science **14**:382-391.

Flemming, H. C. 2011. Microbial biofouling: unsolved problems, insufficient approaches, and possible solutions. Pages 81-109 Biofilm highlights. Springer.

Fonseca, A. C., R. Scott Summers, and M. T. Hernandez. 2001. Comparative measurements of microbial activity in drinking water biofilters. Water Research **35**:3817-3824.

Gilbert, J. J. 1985. Competition between rotifers and Daphnia. Ecology **66**:1943-1950.

Goh, P. S., W. J. Lau, M. H. D. Othman, and A. F. Ismail. 2018. Membrane fouling in desalination and its mitigation strategies. Desalination **425**:130-155.

Greenspan, A. 2011. Evaluation of the Heterotrophic Plate Count Test for Drinking Water Safety: Comparing Culture-based vs. Molecular Methods for Identifying Bacteria.

Haddix, P. L., N. J. Shaw, and M. W. LeChevallier. 2004. Characterization of bioluminescent derivatives of assimilable organic carbon test bacteria. Applied Environmental Microbiology **70**:850-854.

Hamilton, R. D., and O. Holm-Hansen. 1967. Adenosine triphosphate content of marine bacterial. Limnology and Oceanography **12**:319-324.

Hammes, F. 2008. A comparison of AOC methods used by different TECHNEAU partners. TECHNEAU 06. Deliverable 3.310.

Hammes, F., C. Berger, O. Köster, and T. Egli. 2010a. Assessing biological stability of drinking water without disinfectant residuals in a full-scale water supply system. Journal of Water Supply: Research and Technology - Aqua **59**:31-40.

Hammes, F., M. Berney, Y. Wang, M. Vital, O. Köster, and T. Egli. 2008. Flow-cytometric total bacterial cell counts as a descriptive microbiological parameter for drinking water treatment processes. Water Research **42**:269-277.

Hammes, F., F. Goldschmidt, M. Vital, Y. Wang, and T. Egli. 2010b. Measurement and interpretation of microbial adenosine tri-phosphate (ATP) in aquatic environments. Water Research **44**:3915-3923.

Hammes, F. A., and T. Egli. 2005. New method for assimilable organic carbon determination using flow-cytometric enumeration and a natural microbial consortium as inoculum. Environmental Science & Technology **39**:3289-3294.

Henthorne, L., and B. Boysen. 2015. State-of-the-art of reverse osmosis desalination pretreatment. Desalination **356**:129-139.

Hijnen, W. A. M., D. Biraud, E. R. Cornelissen, and D. Van der Kooij. 2009a. Threshold concentration of easily assimilable organic carbon in feedwater for biofouling of spiral-wound membranes. Environmental Science & Technology **43**:4890-4895.

Hijnen, W. A. M., D. Biraud, E. R. Cornelissen, and D. Van Der Kooij. 2009b. Threshold concentration of easily assimilable organic carbon in feedwater for biofouling of spiral-wound membranes. Environmental Science and Technology **43**:4890-4895.

Ho, B. P., M. W. Wu, E. H. Zeiher, and M. Chattoraj. 2004. Method of monitoring biofouling in membrane separation systems.

Hobbie, J. E., R. J. Daley, and S. Jasper. 1977. Use of nuclepore filters for counting bacteria by fluorescence microscopy. Applied and Environmental Microbiology **33**:1225-1228.

Holm-Hansen, O., and C. R. Booth. 1966. The measurement of adenosine triphosphate in the ocean and its ecological significance. Limnology and Oceanography **11**:510-519.

Huber, S. A., A. Balz, M. Abert, and W. Pronk. 2011. Characterisation of aquatic humic and non-humic matter with size-exclusion chromatography – organic carbon detection – organic nitrogen detection (LC-OCD-OND). Water Research **45**:879-885.

Hubley, M. J., B. R. Locke, and T. S. Moerland. 1996. The effects of temperature, pH, and magnesium on the diffusion coefficient of ATP in solutions of physiological ionic strength. Biochimica et Biophysica Acta (BBA) - Bioenergetics **1291**:115-121.

Huck, P. M., P. M. Fedorak, and W. B. Anderson. 1991. Formation and removal of assimilable organic carbon during biological treatment. American Water Works Association:69-80.

Ito, Y., Y. Takahashi, S. Hanada, H. X. Chiura, M. Ijichi, W. Iwasaki, A. Machiyama, T. Kitade, Y. Tanaka, and M. K. a. K. Kogure. 2013. Impact of chemical addition on the establishment of mega-ton per day sized swro desalination plant.*in* YT; Kurihara, Masaru; Kogure, Kazuhiro.

Jacangelo, J. G., N. Voutchkov, M. Badruzzaman, and L. A. Weinrich. 2018. Pretreatment for seawater reverse osmosis: existing plant performance and selection guidance. The Water Research Foundation, The Water Research Foundation.

Jannasch, H. W., and G. E. Jones. 1959. Bacterial populations in sea water as determined by different methods of enumeration. Limnology and Oceanography **4**:128-139.

Jeong, S. 2013. Novel membrane hybrid systems as pretreatment to seawater reverse osmosis. University of Technology, Sydney.

Jeong, S., G. Naidu, and S. Vigneswaran. 2013a. Submerged membrane adsorption bioreactor as a pretreatment in seawater desalination for biofouling control. Bioresource Technology **141**:57-64.

Jeong, S., G. Naidu, S. Vigneswaran, C. H. Ma, and S. A. Rice. 2013b. A rapid bioluminescence-based test of assimilable organic carbon for seawater. Desalination **317**:160-165.

176

Jeong, S., G. Naidu, R. Vollprecht, T. Leiknes, and S. Vigneswaran. 2016. In-depth analyses of organic matters in a full-scale seawater desalination plant and an autopsy of reverse osmosis membrane. Separation and Purification Technology **162**:171-179.

Jeong, S., and S. Vigneswaran. 2015. Practical use of standard pore blocking index as an indicator of biofouling potential in seawater desalination. Desalination **365**:8-14.

Joret, J. C., and Y. Lévi. 1986. Méthode rapide d'évaluation du carbone éliminable des eaux par voies biologiques. La Tribune du CEBEDEAU **510(39)**:3–9.

Kabarty, S. A. 2016. Electrochemical Pretreatment as a Suggested Alternative, More Compatible with Environment and Sustainable Development in RO Desalination System. International Journal of Sustainable and Green Energy **5**:90-102.

Kaplan, L. A., T. L. Bott, and D. J. Reasoner. 1993. Evaluation and simplification of the assimilable organic carbon nutrient bioassay for bacterial growth in drinking water. Applied and Environmental Microbiology **59**:1532-1539.

Karl, D. M. 1980. Cellular nucleotide measurements and applications in microbial ecology. Microbiological Reviews **44**:739-796.

Kemmy, F. A., J. C. Fry, and R. A. Breach. 1989. Development and Operational Implementation of a Modified and Simplified Method for Determination of Assimilable Organic Carbon (AOC) in Drinking Water. Water Science and Technology **21**:155-159.

Khlyntseva, S. V., Y. R. Bazel, A. B. Vishnikin, and V. Andruch. 2009. Methods for the determination of adenosine triphosphate and other adenine nucleotides. Journal of Analytical Chemistry **64**:657-673.

Kim, S. H., C. S. Min, and S. Lee. 2011. Application of dissolved air flotation as pretreatment of seawater desalination. Desalination and Water Treatment **33**:261-266.

Knowles, J. R. 1980. Enzyme-catalyzed phosphoryl transfer reactions. Annual Review of Biochemistry **49**:877-919.

Kujundzic, E., A. C. Fonseca, E. A. Evans, M. Peterson, A. R. Greenberg, and M. Hernandez. 2007. Ultrasonic monitoring of earlystage biofilm growth on polymeric surfaces. Journal of Microbiological Methods **68**:458-467.

LeChevallier, M. W. 1990. Coliform regrowth in drinking water: A Review. American Water Works Association **82**:74-86.

LeChevallier, M. W., N. E. Shaw, L. A. Kaplan, and T. L. Bott. 1993. Development of a rapid assimilable organic carbon method for water. Applied and Environmental Microbiology **59**:1526-1531.

LeChevallier, M. W., N. J. Welch, and D. B. Smith. 1996. Full-scale studies of factors related to coliform regrowth in drinking water. Applied and Environmental Microbiology **62**:2201-2211.

Lee, J., and I. S. Kim. 2011. Microbial community in seawater reverse osmosis and rapid diagnosis of membrane biofouling. Desalination **273**:118-126.

Leenheer, J. A., and J.-P. Croué. 2003. Peer Reviewed: Characterizing Aquatic Dissolved Organic Matter. Environmental Science & Technology **37**:18A-26A.

Li, G. Q., T. Yu, Q. Y. Wu, Y. Lu, and H. Y. Hu. 2017a. Development of an ATP luminescence-based method for assimilable organic carbon determination in reclaimed water. Water Research **123**:345-352.

Li, Q., S. Yu, L. Li, G. Liu, Z. Gu, M. Liu, Z. Liu, Y. Ye, Q. Xia, and L. Ren. 2017b. Microbial Communities Shaped by treatment processes in a drinking water treatment plant and their contribution and threat to drinking water safety. Frontiers in Microbiology **8**:2465-2465.

Liu, G., M. C. Lut, J. Q. Verberk, and J. C. Van Dijk. 2013a. A comparison of additional treatment processes to limit particle accumulation and microbial growth during drinking water distribution. Water Research **47**:2719-2728.

Liu, G., E. J. Van der Mark, J. Q. J. C. Verberk, and J. C. Van Dijk. 2013b. Flow cytometry total cell counts: a field study assessing microbiological water quality and growth in unchlorinated drinking water distribution systems. BioMed Research International **2013**:10.

Ludvigsen, L., H. Albrechtsen, D. B. Ringelberg, F. Ekelund, and T. H. Christensen. 1999. Distribution and Composition of Microbial Populations in a Landfill Leachate Contaminated Aquifer (Grindsted, Denmark). Microbial Ecology **37**:197-207.

Ma, J., A. M. Ibekwe, M. Leddy, C.-H. Yang, and D. E. Crowley. 2012. Assimilable Organic Carbon (AOC) in Soil Water Extracts Using Vibrio harveyi BB721 and Its Implication for Microbial Biomass. PLOS ONE **7**:e28519.

Macdonell, M. T., and M. A. Hood. 1982. Isolation and characterization of ultramicrobacteria from a gulf coast estuary. Applied and Environmental Microbiology **43**:566-571.

Magic-Knezev, A., and D. Van der Kooij. 2004. Optimisation and significance of ATP analysis for measuring active biomass in granular activated carbon filters used in water treatment. Water Research **38**:3971-3979.

Mallick, A. R. 2015. Practical Boiler Operation Engineering and Power Plant. PHI Learning Pvt. Ltd.

Mathias, M., L. Stéphanie, and C. Corinne. 2016. Granular activated carbon filtration plus ultrafiltration as a pretreatment to seawater desalination lines: Impact on water quality and UF fouling. Desalination **383**:1-11.

Matin, A., Z. Khan, S. Zaidi, and M. Boyce. 2011. Biofouling in reverse osmosis membranes for seawater desalination: phenomena and prevention. Desalination **281**:1-16.

Meltzer, T. H., and M. W. Jornitz. 2006. Pharmaceutical Filtration: The Management of Organism Removal. PDA.

Merck Millipore. 2012. Microfiltration membranes for filtration and venting applications.

Montes, M., E. A. Jaensson, A. F. Orozco, D. E. Lewis, and D. B. Corry. 2006. A general method for bead-enhanced quantitation by flow cytometry. Journal of Immunological Methods **317**:45-55.

Munshi, H. A., M. O. Saeed, T. N. Green, A. A. Al-Hamza, M. A. Farooque, and A. A. Ismail. 2005. Impact of uv irradiation on controlling biofouling problems in NF-SWRO desalination process. International Desalination Association (IDA) World Congress. International Desalination Association, Singapore.

Naidu, G., S. Jeong, S. Vigneswaran, and S. A. Rice. 2013. Microbial activity in biofilter used as a pretreatment for seawater desalination. Desalination **309**:254-260.

Nescerecka, A., T. Juhna, and F. Hammes. 2016. Behavior and stability of adenosine triphosphate (ATP) during chlorine disinfection. Water Research **101**:490-497.

Nguyen, T., F. A. Roddick, and L. Fan. 2012a. Biofouling of water treatment membranes: a review of the underlying causes, monitoring techniques and control measures. Membranes (Basel) **2**:804-840.

Nguyen, T., F. A. Roddick, and L. Fan. 2012b. Biofouling of Water Treatment Membranes: A Review of the Underlying Causes, Monitoring Techniques and Control Measures. Membranes 2:804-840.

Nir, T., E. Arkhangelsky, I. Levitsky, and V. Gitis. 2009. Removal of phosphorus from secondary effluents by coagulation and ultrafiltration. Desalination and Water Treatment 8:24-30.

Peleka, E. N., and K. A. Matis. 2008. Application of flotation as a pretreatment process during desalination. Desalination 222:1-8.

Pena N., Del Vigo F., Chesters S. P., Armstrong M. W., Wilson R. , and Fazel M. 2013. A study of the physical and chemical damage on reverse osmosis membranes detected by autopsies. The International Desalination Association World Congress on Desalination and Water Reuse 2013 / Tianjin, China REF: IDAWC/TIAN13-184.

Petry, M., M. A. Sanz, C. Langlais, V. Bonnelye, J.-P. Durand, D. Guevara, W. M. Nardes, and C. H. Saemi. 2007. The El Coloso (Chile) reverse osmosis plant. Desalination 203:141-152.

Piontkovski, S., A. Al-Azri, and K. Al-Hashmi. 2011. Seasonal and interannual variability of chlorophyll-a in the Gulf of Oman compared to the open Arabian Sea regions. International Journal of Remote Sensing 32:7703-7715.

Postgate, J. R. 1969. Chapter XVIII viable counts and viability. Methods in Microbiology 1:611-628.

Poussade, Y., F. Vergnolle, D. Baaklini, N. Pitt, A. Gaid, C. Ventresque, and N. Vigneron-Larosa. 2017. Impact of granular media vs membrane filtration on the pretreatment of SWRO desalination plants The International Desalination Association World Congress São Paulo, Brazil

Prest, E. I., F. Hammes, M. C. M. van Loosdrecht, and J. S. Vrouwenvelder. 2016. Biological stability of drinking water: controlling factors, methods, and challenges. Frontiers in Microbiology 7:45.

Qiu, T. Y., and P. A. Davies. 2015. Concentration polarization model of spiral-wound membrane modules with application to batch-mode RO desalination of brackish water. Desalination 368:36-47.

Quek, S.-B., L. Cheng, and R. Cord-Ruwisch. 2015. Detection of low concentration of assimilable organic carbon in seawater prior to reverse osmosis membrane using microbial electrolysis cell biosensor. Desalination and Water Treatment 55:2885-2890.

Ross, P. S., F. Hammes, M. Dignum, A. Magic-Knezev, B. Hambsch, and L. C. Rietveld. 2013. A comparative study of three different assimilable organic carbon (AOC) methods: results of a round-robin test. Water Science and Technology: Water Supply 13:1024-1033.

Sabiri, N.-E., V. Séchet, P. Jaouen, M. Pontié, A. Massé, and S. Plantier. 2017. Impact of granular filtration on ultrafiltration membrane performance as pre-treatment to seawater desalination in presence of algal blooms. Journal of Water Reuse and Desalination 8:262-277.

Sack, E. L., P. W. Van der Wielen, and D. Van der Kooij. 2010. Utilization of oligo- and polysaccharides at microgram-per-litre levels in freshwater by Flavobacterium Johnsoniae. Journal of Applied Microbiology 108:1430-1440.

Saeki, D., H. Karkhanechi, H. Matsuura, and H. Matsuyama. 2016. Effect of operating conditions on biofouling in reverse osmosis membrane processes: Bacterial adhesion, biofilm formation, and permeate flux decrease. Desalination 378:74-79.

Salinas Rodriguez, S. G., N. Sithole, N. Dhakal, M. Olive, J. C. Schippers, and M. D. Kennedy. 2019. Monitoring particulate fouling of North Sea water with SDI and new ASTM MFI0.45 test. Desalination 454:10-19.

Sarma, Y., A. Al Azri, and S. L. Smith. 2012. Inter-annual Variability of Chlorophyll-a in the Arabian Sea and its Gulfs. International Journal of Marine Science 2.

Saunders, G. W. 1957. Interrelations of dissolved organic matter and phytoplankton. The Botanical Review 23:389-409.

Schippers, J. C., and J. Verdouw. 1980. The modified fouling index, a method of determining the fouling characteristics of water. Desalination 32:137-148.

Schneider, R. P., L. M. Ferreira, P. Binder, E. M. Bejarano, K. P. Góes, E. Slongo, C. R. Machado, and G. M. Z. Rosa. 2005. Dynamics of organic carbon and of bacterial populations in a conventional pretreatment train of a reverse osmosis unit experiencing severe biofouling. Journal of Membrane Science 266:18-29.

She, Q., R. Wang, A. G. Fane, and C. Y. Tang. 2016. Membrane fouling in osmotically driven membrane processes: A review. Journal of Membrane Science **499**:201-233.

Shrestha, A., S. Jeong, S. Vigneswaran, and J. Kandasamy. 2014. Seawater biofiltration pre-treatment system: comparison of filter media performance. Desalination and Water Treatment **52**:6325-6332.

Shutova, Y., B. L. Karna, A. C. Hambly, B. Lau, R. K. Henderson, and P. Le-Clech. 2016. Enhancing organic matter removal in desalination pretreatment systems by application of dissolved air flotation. Desalination **383**:12-21.

Siebel, E., Y. Wang, T. Egli, and F. Hammes. 2008. Correlations between total cell concentration, total adenosine tri-phosphate concentration and heterotrophic plate counts during microbial monitoring of drinking water. Drinking water engineering and science discussions **1**:1-6.

Simon, F. X., Y. Penru, A. R. Guastalli, S. Esplugas, J. Llorens, and S. Baig. 2013a. NOM characterization by LC-OCD in a SWRO desalination line. Desalination and Water Treatment **51**:1776-1780.

Simon, F. X., E. Rudé, J. Llorens, and S. Baig. 2013b. Study on the removal of biodegradable NOM from seawater using biofiltration. Desalination **316**:8-16.

Sintes, E., K. Stoderegger, V. Parada, and G. J. Herndl. 2010. Seasonal dynamics of dissolved organic matter and microbial activity in the coastal North Sea. Aquatic Microbial Ecology **60**:85-95.

Staley, J. T., and A. Konopka. 1985. Measurement of in situ activities of nonphotosynthetic microorganisms in aquatic and terrestrial habitats. Annual Review of Microbiology **39**:321-346.

Stanfield, G., and P. H. Jago. 1987. The development and use of a method for measuring the concentration of assimilable organic carbon in water.

Sung, J. H., M.-S. Chun, and H. J. Choi. 2003. On the behavior of electrokinetic streaming potential during protein filtration with fully and partially retentive nanopores. Journal of Colloid and Interface Science **264**:195-202.

Taverniers, I., M. De Loose, and E. Van Bockstaele. 2004. Trends in quality in the analytical laboratory. II. Analytical method validation and quality assurance. TrAC Trends in Analytical Chemistry **23**:535-552.

Thompson, A. W., and G. van den Engh. 2016. A multi-laser flow cytometry method to measure single cell and population-level relative fluorescence action spectra for the targeted study and isolation of phytoplankton in complex assemblages. Limnology and Oceanography: Methods **14**:39-49.

Vaccaro, R. F., and H. W. Jannasch. 1966. Studies on Heterotrophic Activity in Seawater Based on Glucose Assimilation. Limnology and Oceanography **11**:596-607.

Valavala, R., J. Sohn, J. Han, N. Her, and Y. Yoon. 2011. Pretreatment in reverse osmosis seawater desalination: a short review. Environmental Engineering Research **16**:205-212.

Vallino, J., C. Hopkinson, and J. Hobbie. 1996. Modeling bacterial utilization of dissolved organic matter: optimization replaces Monod growth kinetics. Limnology and Oceanography **41**:1591-1609.

Van der Kooij, D. Albrechtsen, H. , C. Corfitzen, J. Ashworth, I. Parry, F. Enkiri, B. Hambsch, C. Hametner, R. Kloiber, and V. H. 2003. Assessment of the microbial growth support potential of products in contact with drinking water., CPDW Project: European Commission Joint Research Centre.

Van der Kooij, D. 1992. Assimilable organic carbon as an indicator of bacterial regrowth. American Water Works Association **84**:57-65.

Van der Kooij, D., and W. A. M. Hijnen. 1984. Substrate utilization by an Oxalate-consuming Spirillum species in relation to its growth in ozoated water. Applied and Environmental Microbiolgy **Vol. 47, No. 3**:p551- 559.

Van der Kooij, D., and P. W. J. J. Van der Wielen. 2013. Microbial growth in drinking-water supplies: problems, causes, control and research needs. IWA Publishing.

Van der Kooij, D., H. R. Veenendaal, E. J. van der Mark, and M. Dignum. 2017. Assessment of the microbial growth potential of slow sand filtrate with the biomass production potential test in comparison with the assimilable organic carbon method. Water Research.

Van der Kooij, D., A. Visser, and W. A. M. Hijnen. 1982. Determining the concentration of easily assimilable organic carbon in drinking water. American Water Works Association **74**:540-545.

Van der Merwe, R., F. Hammes, S. Lattemann, and G. Amy. 2014. Flow cytometric assessment of microbial abundance in the near-field area of seawater reverse osmosis concentrate discharge. Desalination **343**:208-216.

Van der Wielen, P. W., and D. Van der Kooij. 2010. Effect of water composition, distance and season on the adenosine triphosphate concentration in unchlorinated drinking water in the Netherlands. Water Research **44**:4860-4867.

Van Loosdrecht, M., L. Bereschenko, A. Radu, J. C. Kruithof, C. Picioreanu, M. L. Johns, and H. S. Vrouwenvelder. 2012. New approaches to characterizing and understanding biofouling of spiral wound membrane systems. Water Science and Technology **66**:88-94.

Van Nevel, S., S. Koetzsch, C. R. Proctor, M. D. Besmer, E. I. Prest, J. S. Vrouwenvelder, A. Knezev, N. Boon, and F. Hammes. 2017. Flow cytometric bacterial cell counts challenge conventional heterotrophic plate counts for routine microbiological drinking water monitoring. Water Research **113**:191-206.

Van Slooten, C., T. Wijers, A. G. J. Buma, and L. Peperzak. 2015. Development and testing of a rapid, sensitive ATP assay to detect living organisms in ballast water. Journal of Applied Phycology **27**:2299-2312.

Vang, Ó. K., C. B. Corfitzen, C. Smith, and H.-J. Albrechtsen. 2014. Evaluation of ATP measurements to detect microbial ingress by wastewater and surface water in drinking water. Water Research **64**:309-320.

Velten, S., F. Hammes, M. Boller, and T. Egli. 2007. Rapid and direct estimation of active biomass on granular activated carbon through adenosine tri-phosphate (ATP) determination. Water Research **41**:1973-1983.

Veza, J. M., M. Ortiz, J. J. Sadhwani, J. E. Gonzalez, and F. J. Santana. 2008. Measurement of biofouling in seawater: some practical tests. Desalination **220**:326-334.

Villacorte, L. O. 2014. Algal blooms and membrane based desalination technology. Delft University of Technolog, CRC Press/Balkema.

Villacorte, L. O., Y. Ekowati, H. N. Calix-Ponce, V. Kisielius, J. M. Kleijn, J. S. Vrouwenvelder, J. C. Schippers, and M. D. Kennedy. 2017. Biofouling in capillary and spiral wound membranes facilitated by marine algal bloom. Desalination **424**:74–84.

Visvanathan, C., N. Boonthanon, A. Sathasivan, and V. Jegatheesan. 2003. Pretreatment of seawater for biodegradable organic content removal using membrane bioreactor. Desalination **153**:133-140.

Volk, C. J. 2001. Biodegradable organic matter measurement and bacterial regrowth in potable water. Methods in Enzymology **337**:144-170.

Voutchkov, N. 2010. Considerations for selection of seawater filtration pretreatment system. Desalination **261**:354-364.

Vrouwenvelder, J., S. Manolarakis, H. Veenendaal, and D. Van der Kooij. 2000. Biofouling potential of chemicals used for scale control in RO and NF membranes. Desalination **132**:1-10.

Vrouwenvelder, J., J. Van Paassen, L. Wessels, A. Van Dam, and S. Bakker. 2006. The membrane fouling simulator: a practical tool for fouling prediction and control. Journal of Membrane Science **281**:316-324.

Vrouwenvelder, J. S., J. C. Kruithof, and M. C. Van Loosdrecht. 2010. Integrated approach for biofouling control. Water Science & Technology **62**:2477-2490.

Vrouwenvelder, J. S., S. A. Manolarakis, J. P. Van der Hoek, J. A. M. Van Paassen, W. G. J. Van der Meer, J. M. C. Van Agtmaal, H. D. M. Prummel, J. C. Kruithof, and M. C. M. Van Loosdrecht. 2008. Quantitative biofouling diagnosis in full scale nanofiltration and reverse osmosis installations. Water Research **42**:4856-4868.

Vrouwenvelder, J. S., and D. Van der Kooij. 2001. Diagnosis, prediction and prevention of biofouling of NF and RO membranes. Desalination **139**:65-71.

Vrouwenvelder, J. S., and M. C. M. Van Loosdrecht. 2009. Biofouling of spiral wound membrane systems.

Vrouwenvelder, J. S., M. C. M. Van Loosdrecht, and J. C. Kruithof. 2011. Early warning of biofouling in spiral wound nanofiltration and reverse osmosis membranes. Desalination **265**:206-212.

Vrouwenvelder, J. S., J. A. M. van Paassen, H. C. Folmer, J. A. M. H. Hofman, M. M. Nederlof, and D. van der Kooij. 1998. Biofouling of membranes for drinking water production. Desalination **118**:157-166.

Wang, Q., T. Tao, K. Xin, S. Li, and W. Zhang. 2014. A review research of assimilable organic carbon bioassay. Desalination and Water Treatment **52**:2734-2740.

Wang, Y., F. Hammes, N. Boon, and T. Egli. 2007. Quantification of the filterability of freshwater bacteria through 0.45, 0.22, and 0.1 µm pore size filters and shape-dependent enrichment of filterable bacterial communities. Environmental Science & Technology **41**:7080-7086.

Watanabe, A., R. Ito, and T. Sasa. 1955. Micro-algae as a source of nutrients for daphnids. The Journal of General and Applied Microbiology **1**:137-141.

Webster, J. J., G. J. Hampton, J. T. Wilson, W. C. Ghiorse, and F. R. Leach. 1985. Determination of microbial cell numbers in subsurface samples. Ground Water **23**:17-25.

Weinrich, L., C. N. Haas, and M. W. LeChevallier. 2013. Recent advances in measuring and modeling reverse osmosis membrane fouling in seawater desalination: a review. Journal of Water Reuse and Desalination **3**:85-101.

Weinrich, L., M. LeChevallier, and C. Haas. 2015a. Application of the bioluminescent saltwater assimilable organic carbon test as a tool for identifying and reducing reverse osmosis membrane fouling in desalination. A report submitted to the WateReuse Research Foundation.

Weinrich, L., M. LeChevallier, and C. Haas. 2015b. Application of the bioluminescent saltwater assimilable organic carbon test as a tool for identifying and reducing reverse osmosis membrane fouling in desalination. Water Reuse Research Foundation.

Weinrich, L., M. LeChevallier, and C. N. Haas. 2016. Contribution of assimilable organic carbon to biological fouling in seawater reverse osmosis membrane treatment. Water Research **101**:203-213.

Weinrich, L. A. 2015. The impact of assimilable organic carbon on biological fouling of reverse osmosis membranes in seawater desalination. Drexel University, Drexel University.

Weinrich, L. A., E. Giraldo, and M. W. LeChevallier. 2009. Development and application of a bioluminescence-based test for assimilable organic carbon in reclaimed waters. Applied and Environmental Microbiology **75**:7385-7390.

Weinrich, L. A., P. K. Jjemba, E. Giraldo, and M. W. LeChevallier. 2010. Implications of organic carbon in the deterioration of water quality in reclaimed water distribution systems. Water Research **44**:5367-5375.

Weinrich, L. A., O. D. Schneider, and M. W. LeChevallier. 2011. Bioluminescence-based method for measuring assimilable organic carbon in pretreatment water for reverse osmosis membrane desalination. Applied and Environmental Microbiology **77**:1148-1150.

Werner, P., and B. Hambsch. 1986. Investigations on the growth of bacteria in drinking water. Water Supply **4**:227-232.

White, P. A., J. Kalff, J. B. Rasmussen, and J. M. Gasol. 1991. The effect of temperature and algal biomass on bacterial production and specific growth rate in freshwater and marine habitats. Microbial Ecology **21**:99-118.

WHO. 2006. World Health Organization, Guidelines for drinking-water quality, incorporating first and second addenda.

Wibisono, Y., K. E. El Obied, E. R. Cornelissen, A. J. B. Kemperman, and K. Nijmeijer. 2015. Biofouling removal in spiral-wound nanofiltration elements using two-phase flow cleaning. Journal of Membrane Science **475**:131-146.

Withers, N., and M. Drikas. 1998. Bacterial regrowth potential: quantitative measure by acetate carbon equivalents. Water-Melbourne Then Artarmon **25**:19-23.

LIST OF ACRONYMS

ANOVA	Analysis of variation
AOC	Assimilable organic carbon
ASTM	American Society for Testing and Materials
ASW	Artificial seawater
ATP	Adenosine triphosphate
BDOC	Biodegradable dissolved organic carbon
BFR	Biofilm formation rate
BGP	Bacterial growth potential
BRP	Bacterial regrowth potential
BPP	Biomass production potential
CFU	Colony forming units
CIP	Cleaning in place
CDOC	Chromatography dissolved organic carbon
COD	Chemical oxygen demand
DAF	Dissolved air flotation
DMF	Dual media filter
DMSO	Dimethyl sulfoxide
DOC	Dissolved organic carbon
EBCT	Empty bed contact time
EC	Electrical conductivity
ED	Electro-dialysis
FCM	Flow cytometry
GAC	Granular activated carbon
HPC	Heterotrophic plate count
ICC-FCM	Intact cell concentration measured by flow cytometry
LC-OCD	Liquid chromatography - Organic Carbon Detection
LMW	Low molecular weight
LOD	Limit of detection

mBFR	Biofilm formation rate monitor
MFI	Modified fouling index
MFS	Membrane fouling simulator
n	Number of samples
NOM	Natural organic matter
R2	Regression coefficient
Rho	Spearman regression
RLU	Relative light unit
RO	Reverse osmosis
SDI	Silt density index
SWRO	Seawater reverse osmosis
TCC	Total cell count
TDC	Total direct cell count
TDS	Total dissolved solids
TEP	Transparent exo-polymer particles
TOC	Total organic carbon
UF	Ultrafiltration
UV	Ultraviolet
VC	Variation coefficient

LIST OF TABLES

LIST OF FIGURES

ABOUT THE AUTHOR

Almotasembellah Abushaban

Jul. 2014 – Dec. 2019   PhD researcher at Environmental Engineering and Water Technology department, IHE Delft and at Civil Engineering and Geoscience Faculty, TU Delft, Netherlands.

Oct. 2012 – Jun. 2014   Master of science in Water Supply Engineering at UNESCO-IHE, Delft, Netherlands.

Sep. 2011 – Oct. 2012   Teaching Assistant at Environmental and Civil Engineering Department, The Islamic University of Gaza, Palestine.

Jan. 2012 – Oct. 2012   Coordinator of the 4th International Engineering Conference at the Engineering Faculty of the Islamic University of Gaza, Palestine.

Oct. 2010 – Oct. 2012   Program officer for the Gulf Cooperation Council projects in Gaza at the Engineering and Management Consulting Centre (EMCC), Palestine.

25 Sep. 1988   Born in Kuwait, Kuwait

- ## Awards and fellowships

1. **The Innovation Award of The International Desalination Association (IDA)** at the world congress 2017 in Sao Paulo, Brazil (October 2017).

2. **MSc Scholarship** from the Middle East Desalination Research Center (MEDRC), in Muscat, Oman to study MSc in Water Supply Engineering at UNESCO-IHE, Delft, Netherlands (October 2012-April 2014).

3. **Best scientific research award in the Engineering Faculty** from the Islamic university of Gaza, Palestine (April 2012).

4. **Environmental Creativity Award** – First rank from the Ministry of youth and sports, Palestine (February 2012).

5. **BSc Scholarship** from Hani Qaddumi Scholarship in Amman, Jordan to study Civil Engineering at the IUG, Palestine (September 2008 -December 2010)

6. **Best student fellowship** to participate in several courses related to leadership and management, project management, advance English at the Islamic University of Gaza (September 2009- June 2010).

7. **Honour list and Honour Certificates** (Seven times): at the Islamic University of Gaza (September 2006 – December 2010).

- **Journal publications**

1. Abushaban A., Salinas-Rodriguez S.G., Dhakal N., Schippers J.C., and Kennedy M.D. (2019) Assessing pre-treatment and seawater reverse osmosis performance using an ATP-based bacterial growth potential method. *Desalination.* v(467) p 210–218.

2. Abushaban A., Salinas-Rodriguez S.G., Mangal M.N., Mondal S., Goueli S.A., Knezev A., Vrouwenvelder J.S., Schippers J.C., and Kennedy M.D. (2019) ATP measurement in seawater reverse osmosis systems: eliminating seawater matrix effects using a filtration-based method. *Desalination.* v(453) p 1–9.

3. Abushaban A., Mangal M.N., Salinas-Rodriguez S.G., Nnebuoa C., Mondal S., Goueli S.A., Schippers J.C., and Kennedy M.D. (2017) Direct measurement of ATP in seawater and application of ATP to monitor bacterial growth potential in SWRO pre-treatment systems. *Desalination and Water Treatment*, v (99), p91-101.

4. Salinas-Rodriguez S.G., Prabowo A., Abushaban A., Schippers J.C., and Kennedy M.D., (2016) Pre-coating of outside-inside ultrafiltration systems with iron hydroxide particles to control non-backwashable fouling. *Desalination and Water Treatment*, v (57), p26730-26740.

5. Hamad J.T., Eshtawi T.A., Abushaban A., and Habboub M.O. (2012) Modeling the impact of land-use change on Water Budget of Gaza Strip- *Journal of Water Resource and Protection*, Published Online June 2012, 4, 325-333.

- ## Journal articles submitted/ in preparation

1. Abushaban A. , Salinas-Rodriguez S. G., Kapala M., Pastorelli D., Schippers J. C., Mondal S., Goueli S., Kennedy M.D. (2019). Correlating bacterial growth potential measurement to real time fouling development in full-scale SWRO. Submitted to Desalination.

2. Abushaban A., Salinas-Rodriguez S.G., Pastorelli D., Goueli S.A., Schippers J.C., and Kennedy M.D. (2019) Application of ATP and BGP methods to monitor media filtration and dissolved air flotation pre-treatment systems. In preparation

3. Abushaban A., Salinas-Rodriguez S.G., Schippers J.C., and Kennedy M.D. Monitoring biological degradation using dissolved oxygen measurements. In preparation.

- ## Conference publications

1. Abushaban A., Salinas-Rodriguez S.G., Pastorelli D., Saul B., Schippers J.C., and Kennedy M.D. (2019) Monitoring fouling potential along the pre-treatment of an seawater reverse osmosis desalination plant, *The International Desalination Association World Congress* – Dubai- UAE (REF: IDA19WC- Abushaban).

2. Abushaban A., Salinas-Rodriguez S.G., Mondal S., Goueli S.A., Schippers J.C., and Kennedy M.D. (2017) A new method of assessing bacterial growth in SWRO systems: Method development and applications, *The International Desalination Association World Congress* – São Paulo, Brazil (REF: IDA17WC- 58031_ Abushaban).

3. Mondal S., Abushaban A., Mangal M.N., Salinas-Rodriguez, S.G., Goueli S., and Kennedy M.D (2017) Development and Application of New Methods to Measure Bacterial Activity and Nutrients in Seawater Reverse Osmosis (SWRO). *AMTA/AWWA Membrane Technology Conference*, Long Beach, CA. Ref. (9676-DP1950).

4. Abushaban A., Salinas-Rodriguez S.G., Mondal S., Goueli S., Schippers J.C., and Kennedy M.D. (2018) Monitoring Adenosine Triphosphate and bacterial

regrowth potential in Seawater Reverse Osmosis Plants, presented in *12th AEDyR International Congress, Toledo Congress Centre El Greco*, Spain, October 23– 25, 2018.

• Conference presentations

1. Abushaban A., Villacorte L., Salinas-Rodriguez S.G., Schippers J.C., and Kennedy M.D. (2015) *Measuring Bacterial Adenosine Triphosphate (ATP) in Seawater: method development and application in seawater reverse osmosis*, Presented in EuroMed— Desalination for Clean Water and Energy, Palermo, Italy, May 10-14, 2015.

2. Abushaban A., Salinas-Rodriguez S.G., Schippers J.C., and Kennedy M.D. (2015) *Assessing bacterial growth in seawater reverse osmosis: A new method for measuring bacterial adenosine triphosphate (ATP)*, presented in the 8th International Desalination Workshop (IDW 2015), Jeju Island, Korea, November 18–21, 2015.

3. Abushaban A., Mangal M.N., Salinas-Rodriguez S.G., Mondal S., Goueli S., Schippers J.C., and Kennedy M.D. (2016) *Measuring Adenosine Triphoshate (ATP) and Assimiable Organic Carbon (AOC) in Seawater Reverse Osmosis (SWRO) Feed water*, presented in EDS conference - Desalination for the Environment Clean Water and Energy, Rome, Italy, May 22–26, 2016.

4. Abushaban A., Ramrattan L., Salinas-Rodriguez S.G., Yangali-Quintanilla V., Blankert B., Schippers J.C., and Kennedy M.D. (2016) *Induction time of phosphonate antiscalant in the absence and presence of RO membrane surface and feed spacer*, presented in EDS conference - Desalination for the Environment Clean Water and Energy, Rome, Italy, May 22–26, 2016.

5. Abushaban A., Salinas-Rodriguez S.G., Mondal S., Goueli S., Schippers J.C., and Kennedy M.D. (2016) *Assessing Bacterial Growth In Seawater RO Systems: A New Method For Measuring Bacterial ATP*, presented in IWA World Water Congress and Exhibition, Brisbane, Australia, October 9–13, 2016.

201

6. Abushaban A., Salinas-Rodriguez S.G., Mondal S., Goueli S., Schippers J.C., and Kennedy M.D. (2016) *Assimilable Organic Carbon in seawater based on Bacterial Adenosine Triphosphate: Method development and application*, presented in the 9th International Desalination Workshop, Yas Island, Abu Dhabi, UAE, November 13–15, 2016,

7. Abushaban A., Salinas-Rodriguez S.G., Mondal S., Goueli S., Schippers J.C., and Kennedy M.D. (2017) *A new method of assessing bacterial growth in SWRO systems: Method development and applications*, presented in Membranes in the Production of Drinking and Industrial Water (MDIW), Leeuwarden, The Netherlands, February 6–8, 2017.

8. Mondal S., Abushaban A., Mangal M.N., Salinas-Rodriguez, S.G., Goueli S., and Kennedy M.D (2017) *Development and Application of New Methods to Measure Bacterial Activity and Nutrients in Seawater Reverse Osmosis (SWRO)*, presented in Membrane Technology Conference and Exposition AMTA/AWWA, Long Beach, USA, February 13–17, 2017.

9. Abushaban A., Salinas-Rodriguez S.G., Mondal S., Goueli S., Schippers J.C., and Kennedy M.D. (2017) *Assessing biofouling in seawater reverse osmosis systems: Method development and applications*, presented in the 3rd International Conference on Desalination Using Membrane Technology in Las Palmas, Gran Canaria, Spain, April 2-5, 2017.

10. Abushaban A., Salinas-Rodriguez S.G., Mondal S., Goueli S., Schippers J.C., and Kennedy M.D. (2017) *Measuring ATP and AOC in Seawater Reverse Osmosis Plants*, presented in EuroMed 2017 — Desalination for Clean Water and Energy: Cooperation around the World, Tel Aviv, Israel, May 9–12, 2017.

11. Abushaban A., Salinas-Rodriguez S.G., Mondal S., Goueli S., Schippers J.C., and Kennedy M.D. (2017*) Development and application of a new method for assessing bacterial growth in seawater reverse osmosis*, presented in 11th International Congress on Membranes and Membrane Processes (ICOM), San Francisco, CA, USA, July 29 – August 4, 2017.

12. Abushaban A., Salinas-Rodriguez S.G., Pastorelli D., Mondal S., Schippers J.C., and Kennedy M.D. (2018) *Monitoring adenosine triphosphate and bacterial regrowth potential along the pre-treatment of an Seawater Reverse Osmosis Plant*, presented in Desalination for the Environment: Clean Water and Energy, Athens, Greece, September 3–6, 2018.

13. Abushaban A., Salinas-Rodriguez S.G., Mondal S., Goueli S., Schippers J.C., and Kennedy M.D. (2018) *Monitoring Adenosine Triphosphate and bacterial regrowth potential in Seawater Reverse Osmosis Plants*, presented in 12th AEDyR International Congress, Toledo Congress Centre El Greco, Spain, October 23–25, 2018.

14. Abushaban A., Salinas-Rodriguez S.G., Pastorelli D., Saul B., Schippers J.C., and Kennedy M.D. (2019) *Monitoring the pre-treatment of a full-scale seawater reverse osmosis desalination plant,* presented in the International Desalination Association World Congress 2019, Dubai, UAE, October 20-24, 2019.

SENSE

Netherlands Research School for the
Socio-Economic and Natural Sciences of the Environment

D I P L O M A

For specialised PhD training

The Netherlands Research School for the
Socio-Economic and Natural Sciences of the Environment
(SENSE) declares that

Almotasembellah M. J. Abushaban

born on 25 September 1988 in Kuwait, Kuwait

has successfully fulfilled all requirements of the
Educational Programme of SENSE.

Delft, 3 December 2019

The Chairman of the SENSE board

Prof. dr. Martin Wassen

the SENSE Director of Education

Dr. Ad van Dommelen

The SENSE Research School has been accredited by the Royal Netherlands Academy of Arts and Sciences (KNAW)

K O N I N K L I J K E N E D E R L A N D S E
A K A D E M I E V A N W E T E N S C H A P P E N

The SENSE Research School declares that Almotasembellah M. J. Abushaban has successfully fulfilled all requirements of the Educational PhD Programme of SENSE with a work load of 41.3 EC, including the following activities:

<u>SENSE PhD Courses</u>

o Environmental research in context (2016)
o Research in context activity: 'Preparing, conducting and reporting on a research campaign at Barka II desalination plant in Oman (20 April - 4 May 2017)'
o Basic Statistics (2017)

<u>External training at a foreign research institute</u>

o Designing the pre-treatment membrane systems for RO, Pentair X-flow Academy, Enschede, The Netherlands (2015)

<u>Management and Didactic Skills Training</u>

o Supervising six MSc students with thesis (2014-2019)
o Supervising BSc student with thesis entitled 'Further development of Adenosine Triphosphate and bacterial growth potential methods to monitor biofouling in Seawater reverses osmosis' (2018)
o Teaching staff members of Water authority of Jordan, Aqaba water company and Marine Science Station of Jordan University (2019)
o Water Advisory Board meeting, Promega, USA (2016)
o Member of PhD Association Board IHE Delft (2015-2016)
o Organising PhD week, IHE Delft (2016)

<u>Selection of Oral Presentations</u>

o *Measuring Bacterial Adenosine Triphosphate (ATP) in Seawater: method development and application in seawater reverse osmosis.* EuroMed-Desalination for Clean Water and Energy, 10-14 May 2015, Palermo, Italy
o *Assessing bacterial growth in seawater RO systems: a new method for measuring bacterial ATP*, IWA World Water Congress and Exhibition, 9-13 October 2016, Brisbane, Australia
o *Assessing biofouling in seawater reverse osmosis systems: Method development and applications.* The 3rd International Conference on Desalination Using Membrane Technology, 2-5 April 2017, Las Palmas, Gran Canaria, Spain
o *Monitoring adenosine triphosphate and bacterial regrowth potential along the pre-treatment of Seawater Reverse Osmosis Plant.* Desalination for the Environment: Clean Water and Energy, 3-6 September 2018 Athens, Greece

SENSE Coordinator PhD Education

Dr. Peter Vermeulen